THE
FIELD GUIDE TO
GEOLOGY

New Edition

THE
FIELD GUIDE TO
GEOLOGY

New Edition

David Lambert
and the Diagram Group

☑Checkmark Books®
An imprint of Infobase Publishing

The Field Guide to Geology, New Edition

Copyright © 2007 by Diagram Visual Information Ltd.

Editors	Annabel Else, Denis Kennedy, Gordon Lee, Jamie Stokes
Design	Richard Hummerstone, Arthur Lockwood
Artists	Joe Bonello, Alastair Burnside, Richard Czapnik, Brian Hewson, Lee Lawrence, Chris Logan, Paul McCauley, Philip Patenall, Micky Pledge, Philip Richardson, Jane Robertson, Graham Rosewarne, Debbie Skinner
Indexer	Martin Hargreaves

Checkmark Books
An imprint of Infobase Publishing
132 West 31st Street
New York NY 10001

ISBN-10: 0-8160-6510-1
ISBN-13: 978-0-8160-6510-3

Library of Congress Cataloging-in-Publication Data

Lambert, David, 1932–
   The field guide to geology / David Lambert and the
Diagram Group. — [New ed.]
      p. cm.
   Rev. ed. of: The field guide to geology. Updated ed. c1998.
   Includes bibliographical references and index.
   ISBN 0-8160-6509-8 (acid-free paper) — ISBN 0-8160-6510-1
   (pbk.: acid-free paper) 1. Geology. I. Diagram Group.
II. Title.

   QE28.L22 2006
   550--dc22   2006048533

Checkmark Books are available at special discounts when purchased in bulk quantities for businesses, associations, institutions, or sales promotions. Please call our Special Sales Department in New York at 212/967-8800 or 800/322-8755.

You can find Facts On File on the World Wide Web at http://www.factsonfile.com

Printed in the United States of America

EB DIAG 10 9 8 7 6 5

Consultancy panel

INTRODUCTION

This book provides a concise yet comprehensive key to the constituents and processes that forged our planet. Hundreds of illustrations—large, clearly labelled diagrams, "field guides," and maps—help readers grasp important concepts visually. Images and text are integrated to explain scientific terms by clear illustration. Together, text and pictures offer an up-to-date guide for all, from the inquiring eleven-year-old to the budding Earth scientist.

There are fourteen chapters. Each has a brief explanatory introduction, followed by related topics:

Chapter 1, Sizing Up Earth, gives an overview of our planet's origins and raw materials.

Chapter 2, The Restless Crust, explains the astonishing processes that shape and reshape continents and oceans, and recycle rock.

Chapter 3, Fiery Rocks, deals with igneous rocks—rocks formed from molten matter, the stuff from which all rock derives.

Chapter 4, Rocks from Scraps, covers sedimentary rocks—those mostly formed from reconstituted pieces of igneous or other rocks.

Chapter 5, Deformed and Altered Rocks, investigates processes that displace crustal rocks and change one kind of rock into another.

Chapter 6, Crumbling Rocks, examines the role of the weather in destroying rocks, building soil, and sculpting slopes.

Chapter 7, How Rivers Shape the Land, traces the role of rivers in carving valleys, shaping plains and deltas, and creating lakes.

Chapter 8, The Work of the Sea, tells how the sea builds and destroys land.

Chapter 9, The Work of Ice and Air, explains how glaciers and ice sheets mold cold lands, and how winds sculpt desert landscapes.

Chapter 10, Change through the Ages, shows how geologists read the record of Earth's history in rocks, and what the oldest rocks reveal.

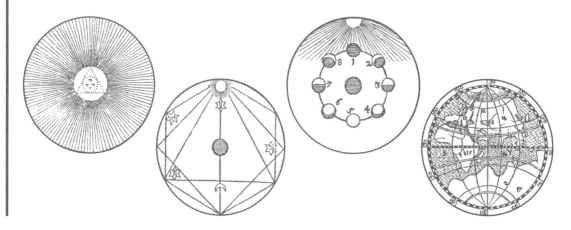

Chapter 11, The Last 543 Million Years, covers the great changes that have formed, reformed, and repositioned continents since rocks acquired a detailed fossil record.

Chapter 12, Rocks and Man, describes how geologists find and exploit useful rocks and minerals.

Chapter 13, Monitoring Earth, is an entirely new section that explains how new technology has made it possible for planet scientists to measure the behavior of Earth's complex systems directly in real time. It encompasses satellite technology, acoustics, and seismic monitoring.

Chapter 14, Who and Where? looks at the lives and discoveries of great scientists who have changed the way we think about Earth. It includes listings for the great geological museums and collections of the world as well as Web sites and associations in the field.

Since this volume was first published, new technologies have revolutionized the way planet scientists now investigate Earth's systems. Technologies such as remote satellite sensing have also allowed scientists to recognize the interconnectedness of Earth systems that were previously studied in isolation. Today, geologists must appreciate the lessons of marine science and atmospheric research to understand their own discipline properly. These advances are addressed in Chapter 13. The biographies of great scientists have also been updated and added to, as have the listings of geological museums and collections. Web site references are also included for the first time.

Middle and high-school students will benefit from an appreciation of the global viewpoint that geologists and other planetary scientists are increasingly taking. They will also receive a basic grounding in the forms of technology that these scientists now rely on to monitor the ever-changing Earth.

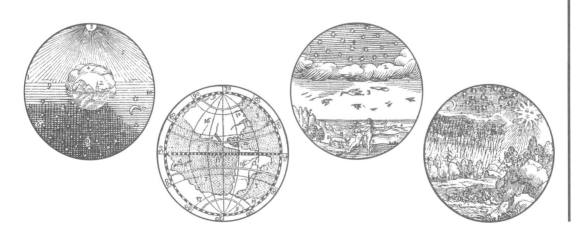

CONTENTS

1 Chapter 1
SIZING UP EARTH

2 Chapter 2
THE RESTLESS CRUST

3 Chapter 3
FIERY ROCKS

4 Chapter 4
ROCKS FROM SCRAPS

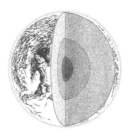
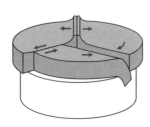

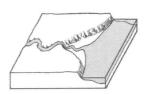
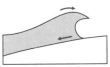

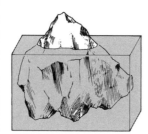
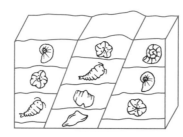

12 Chapter 12
ROCKS AND MAN

13 Chapter 13
MONITORING EARTH

14 Chapter 14
WHO AND WHERE?

CHAPTER 1 SIZING UP EARTH

This book begins by putting our planet in its universal context. We see how matter, stars, and the solar system evolved and how Earth acquired its layered structure and slightly bulging shape. There is a brief overview of elements, minerals, and rocks—Earth's building blocks. The chapter ends with a look at the forces that keep the sea and air in motion, disturb the crust, and turn Earth into a mighty dynamo.

A selection of minerals and their crystalline form. (Engraving originally published in *The Iconographic Encyclopaedia of Science, Literature and Art*, 1851)

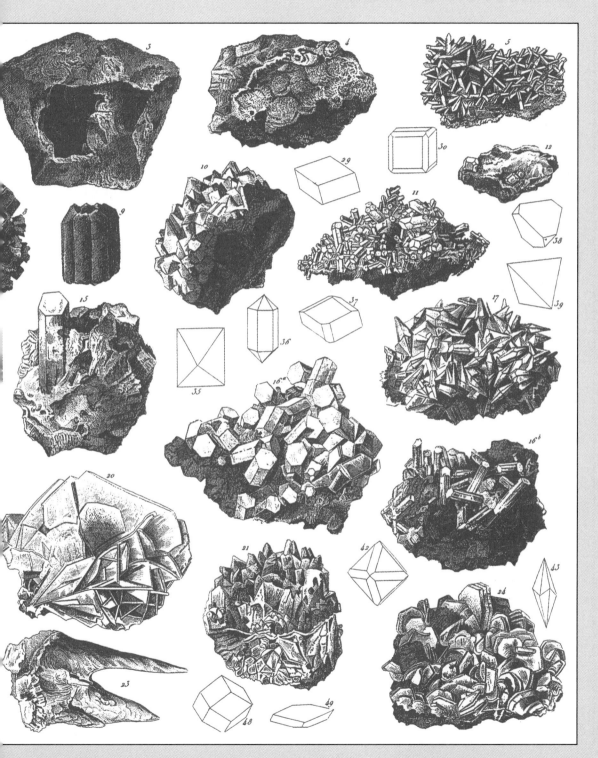

13

EARTH IN SPACE

Sun and planets
Numbered items show
(below) relative sizes of and
(bottom) gaps between the
Sun and planets. Asteroids
(minor planets) lie between 5
and 6.

1 Sun	**6** Jupiter
2 Mercury	**7** Saturn
3 Venus	**8** Uranus
4 Earth	**9** Neptune
5 Mars	**10** Pluto

Earth is a rocky, spinning ball–one of nine planets and many lesser bodies (moons, asteroids, and comets) orbiting a star (the Sun). All of these together constitute our solar system.

Earth is tiny compared to the four largest planets. But our solar system's largest and most influential body is the Sun, a glowing ball of gases a million times the volume of Earth and far bigger than all the other objects orbiting the Sun. The Sun's immense gravitational force prevents the entities around it from flying outward into space. And its electromagnetic radiations produce the heat and light that help make life possible on Earth—the third nearest planet to the Sun. Most planets closer in or farther out appear too hot or cold for life.

Earth's behavior and that of its Moon determine time on Earth. Like all planets, Earth spins on a central axis with imaginary ends at the poles. Each rotation of about 24 hours produces day and night.

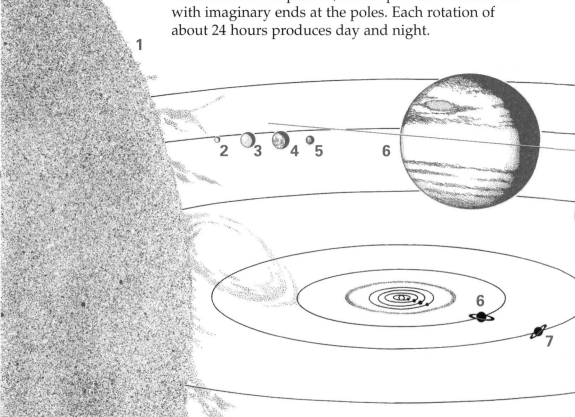

About once a month the Moon completes one revolution around Earth. Earth itself completes one orbit of the Sun in about 365 days—an Earth year. Because Earth orbits in a tilted attitude, sunlight beams down directly upon its Northern and Southern Hemispheres at different times of year, creating seasons.

Immense distances separate Earth from other bodies in space. From Earth to the Moon is about 1.25 light-seconds, the distance covered in that time by light, which travels at 186,000 miles per second (300,000 kmps). From Earth to the Sun is 8 light-minutes; the solar system is 11 light-hours across; from Earth to the nearest star beyond the Sun is 4 light-years. Our solar system plus dust, gas, and 100 billion stars (some certainly with solar systems of their own) comprise our galaxy—a flattened disk 80,000 light-years across. At least 10 billion galaxies are scattered throughout the universe.

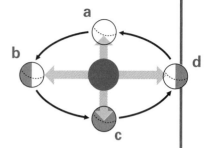

The seasons
Earth's tilt brings the midday sun overhead north or south of the equator at different times of year, creating seasons.
a March 21: Sun over the equator
b June 21: Sun over the tropic of Cancer
c September 23: Sun over the equator
d December 21: Sun over the tropic of Capricorn

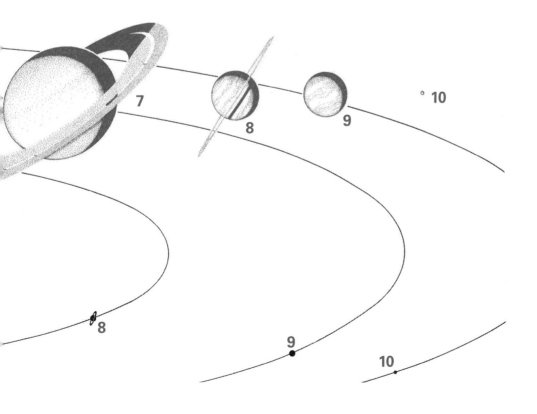

© DIAGRAM

HOW EVERYTHING BEGAN

A

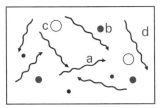

B

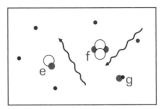

The first atoms *(above)*
A Post big-bang radiation (**a**) and subatomic particles: protons (**b**), neutrons (**c**), and electrons (**d**)
B Subatomic particles combined as nuclei of deuterium ("heavy hydrogen") (**e**), helium (**f**), and hydrogen (**g**)

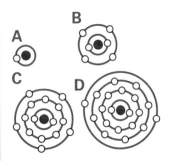

Building bigger atoms
"Star factories" forged heavier, more complex atoms, indicated by the number of electrons orbiting their nuclei.
A Hydrogen
B Carbon
C Phosphorus
D Calcium

Earth's origins lie in the creation of the universe. Just how this came about remains unclear, but many scientists accept some version of the big bang theory, which goes like this: at first all energy and matter (then only subatomic particles) were closely concentrated. About 14 billion years ago a vast explosion scattered everything throughout space. Star studies prove that the universe is still expanding, and background radiation hints at its initial heat.

The big bang sparked off processes that produced atoms that formed different elements—the chemically indivisible building blocks of stars and planets. More than 90 elements occur on Earth alone.

Hydrogen and helium, the lightest, most abundant elements, would have begun forming as subatomic particles within minutes of the big bang. Hundreds of millions of years later, condensing clouds of hydrogen formed galaxies. Here mighty blobs of gas contracted under gravitation. This process warmed the gas and

Rare and common elements
Elements made of large, complex atoms tend to be scarcer than those made of small, simple atoms.
a Logarithmic scale of universal abundance (relative to 1 million atoms of silicon)
b Atomic number (number of electrons per atom)

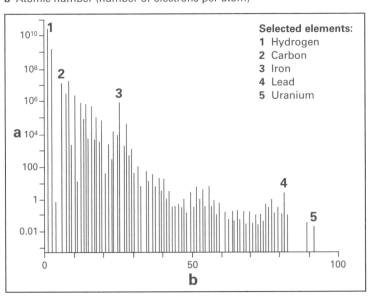

Selected elements:
1 Hydrogen
2 Carbon
3 Iron
4 Lead
5 Uranium

triggered nuclear reactions. These converted hydrogen to helium and gave off energy, including light. Thus blobs of gas evolved into stars.

Stars that had used up all their hydrogen started "burning" helium and swelling into red giants. Inside their nuclear furnaces, atomic evolution sped up. First, helium nuclei combined to form carbon atoms. Next, carbon gave rise to heavier elements, such as neon, nitrogen, and oxygen. Then came still heavier elements, including iron. From red giants such substances escaped into space.

Meanwhile massive "burned out" stars became supernovae, ending in immense explosions that forged heavy elements such as uranium and gold, and hurled them into interstellar space.

New stars acquired these preformed elements. They make up only one percent of the Sun's mass. But much higher percentages occur on planets like Earth. The next pages explain how the Sun and planets formed.

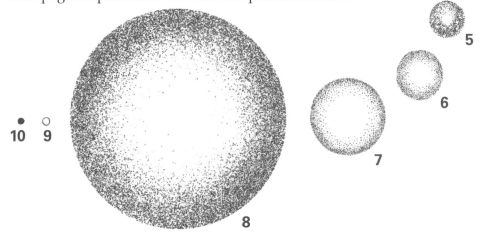

Life of a star (above)
Many stars go through the stages pictured on this page.

1 Gases make up a formless mass called a nebula.
2 Gravitation makes the nebula contract to a regular shape.
3 Contraction heats the center of the nebula until nuclear reactions turn hydrogen to helium and it becomes a star.

4 The star completes its main contraction and starts a long, stable phase of energy output as a so-called yellow dwarf.
5 The star's hydrogen exhausted, its core shrinks; its surface expands and cools.
6 The star grows brighter and expands into a red giant.
7 The star burns helium, heats up, and grows enormously.

8 The expanding star may reach 400 times its former size.
9 Its nuclear energy used up, the star collapses to a tiny, dense, dim white dwarf.
10 The star ends as a black dwarf, no longer yielding heat or light.

© DIAGRAM

BIRTH OF THE SOLAR SYSTEM

Cosmologists believe that the universe was more than 10 billion years old when our solar system formed 4.6 billion years ago. Its raw material was a cloud of dust and gas in one spiral arm of our galaxy. Somehow that cloud became the Sun—a central star containing 99.8 percent of the solar system—and the planets, satellites lying roughly in one plane, most spinning in the same direction, that orbit the Sun.

Scientists have produced various theories to account for this arrangement. Here are four. (The fourth now seems most likely to be right.)

1 Nebular hypothesis A spinning, shrinking cloud of gas and dust produced the Sun and threw off rings, which condensed to form planets and their moons.

2 Tidal theory A passing star pulled a tongue of matter from the Sun. The tongue split into drops, then each drop became a planet.

3 Exploding supernova Our Sun supposedly had a companion star that exploded, leaving scattered debris that gave rise to planets.

1 Nebular hypothesis
Planets form from rings of gas shed by the spinning center of a gas cloud.

2 Tidal theory
A A passing star drags a gaseous tongue from the Sun.
B The tongue breaks into drops that condense into planets.

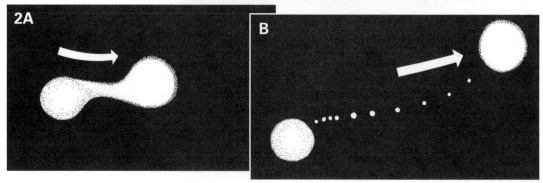

4 Accretion theory Shock waves from supernovae made a cold cloud of gas and dust condense into smaller, denser clouds where complex molecules including methane formed. A collapsing portion of one cloud became a spinning disk of helium and hydrogen with 1–2 percent of heavier elements. Gravitational attraction produced the Sun and the giant gas-rich planets. Meanwhile, dust grains stuck together and created lumps attracting smaller debris and growing into so-called planetesimals 60–600 miles (100–1,000 km) across. After 100 million years or so, coalescing swarms of planetesimals formed Earth and Earth-like planets—largely stripped of hydrogen and helium gases by a stream of solar particles.

3 Exploding supernova
The Sun's companion star explodes, producing debris that forms planets.

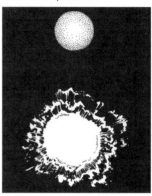

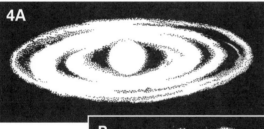

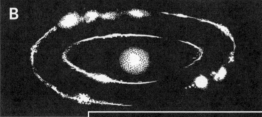

4 Accretion theory
A A cold gas cloud shrinks, and contraction forms a central star (the Sun).
B Condensing masses of gas and dust begin to form planets.
C The contracting Sun grows hot enough to start a nuclear chain reaction.
D At least nine planets now revolve in regular orbits around the brightly shining Sun.

© DIAGRAM

OUR LAYERED PLANET

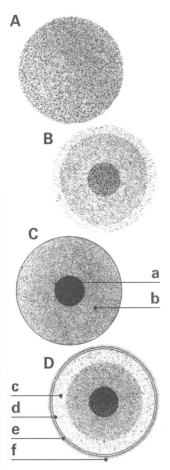

Sorting out Earth
Our planet might have formed in these four stages.
A Cloud of unsorted particles
B Particles sorted, the densest in the middle
C Primeval planet with:
a Dense iron and nickel core
b Material, as in carbonaceous chondrite meteorites
D Outer chondrite melted, yielding:
c Relatively dense mantle
d Primitive crust
e Ocean
f Early atmosphere

Accumulating mini-planets, dust, and gases formed Earth about 4.6 billion years ago. Compression caused by gravitation produced immense internal heat and pressure. Meanwhile gravity was sorting out Earth's ingredients. Heavy elements tended to be drawn into the middle. Lighter elements and compounds collected near the surface. The lightest elements included an outer film of gases. This poisonous primeval atmosphere gave way to air in time, and Earth's molten surface cooled and hardened, but its interior remains intensely hot.

Such processes produced Earth's major layers:

1 Atmosphere An invisible film of gases (now mainly nitrogen and oxygen) that scientists have divided from the base up into the following areas: troposphere, stratosphere, mesosphere, ionosphere, thermosphere, exosphere, and magnetosphere.

2 Crust Earth's rocky outer surface is 4 miles (6 km) thick under oceans and up to 40 miles (64 km) thick under mountain ranges. The crust consists of relatively "light" rocks—largely granite below continents and basalt below oceans. Oceans cover about 71 percent of the crust.

3 Mantle A layer of rocks 1,800 miles (2,900 km) thick between the crust and outer core. Parts are semi-molten and evidently flow in sluggish currents. Mantle rock is more dense than crustal rock.

4 Outer core A layer of dense molten rocks 1,400 miles (2,240 km) thick between mantle and inner core. It may be mainly iron and nickel with some silicon.

5 Inner core A solid ball 1,540 miles (2,440 km) across. Intense pressure prevents it from liquefying despite a temperature of 3,700°C. The inner core may be mainly iron and nickel.

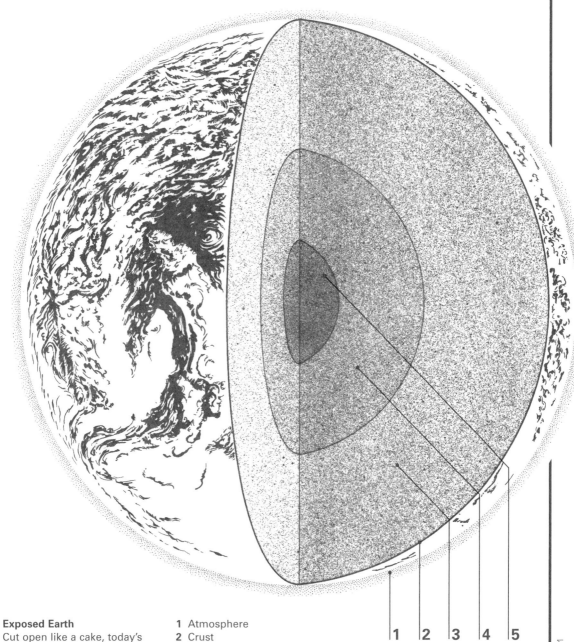

Exposed Earth
Cut open like a cake, today's
Earth would show this
"onion" structure. Scientists
learn about its inner layers by
studying how these interfere
with waves set off by
earthquakes.

1 Atmosphere
2 Crust
3 Mantle
4 Outer core
5 Inner core

1 2 3 4 5

© DIAGRAM

EARTH'S SIZE AND SHAPE

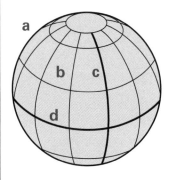

Earth measurements *(above)*
a Mass: 5.98 x 10²⁴ metric
   tons
b Surface area: 196,935
   million square miles
   (510.66 billion km²)
c Polar circumference:
   24,859.7 miles (40,008 km)
d Equatorial circumference:
   24,901.5 miles (40,075 km)

Instruments, including artificial satellites, have helped scientists work out Earth's size, shape, and other features. We now know ours is one of the smallest, lightest planets of the solar system—four others far exceed its mass and volume—but no planet has a greater density (5.5 times that of water).

Careful measurements prove that the ball-like Earth is not in fact a sphere. It measures 24,902 miles (40,075 km) around the equator, but only 24,860 miles (40,008 km) around the poles. So our planet bulges slightly at the equator and is slightly flattened at the poles. Centrifugal force created by Earth's spin produced this shape—an oblate spheroid. Even that description oversimplifies, for Earth is very slightly pear-shaped.

Scientists accordingly use the term *geoid* ("earth shaped") to describe Earth's hypothetical, mean sea-

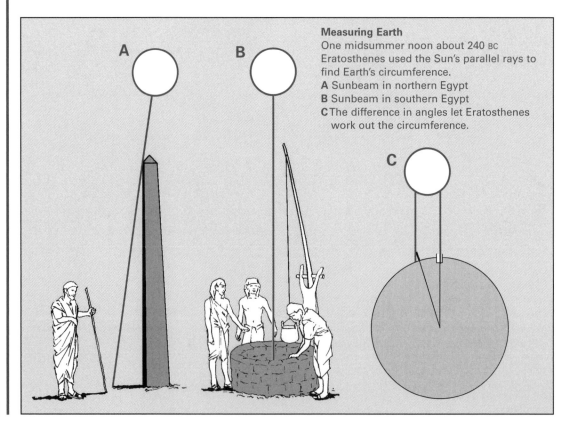

Measuring Earth
One midsummer noon about 240 BC Eratosthenes used the Sun's parallel rays to find Earth's circumference.
A Sunbeam in northern Egypt
B Sunbeam in southern Egypt
C The difference in angles let Eratosthenes work out the circumference.

A

B

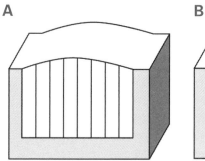

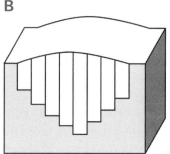

Ideas about isostasy (*left*)
A In 1856 John Pratt argued that high and low mountains were blocks of rocks of different density floating above the same base level.
B Sir George Airy thought that different heights showed blocks—of the same density but different thickness—floating at different depths. Airy's view prevailed.

level surface—ignoring wrinkles formed by mountain chains and ocean floors. Geoid measurement involves taking sea-level gravity readings by gravimeter and studying "kinks" in the orbits of artificial satellites. Both reveal so-called gravity anomalies that reflect local differences in mass in Earth's crust and mantle. Such differences account for vast but slight dips and bumps in the geoid's surface.

Gravity anomalies also reinforce the theory of isostasy—a state of balance in Earth's crust where continents of light material float on a denser substance into which deep continental "roots" project like the underwater mass of floating icebergs.

The global geoid (*below*)
This world map shows the global geoid's surface areas (**a**) above and (**b**) below the surface of a true ellipsoid, as deduced from surface gravity and artificial satellites. The geoid's highest "bumps" are little more than 558 feet (170 m) above its lowest depressions.

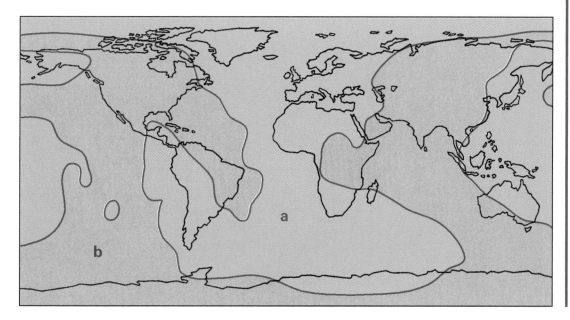

© DIAGRAM

EARTH'S BUILDING BLOCKS 1

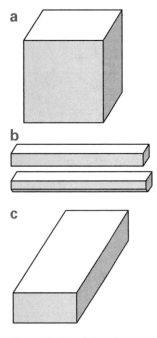

Earth's building blocks are elements and minerals. Of our planet's 92 naturally occurring elements, eight account for 98 percent of the weight of Earth's crust—the rocky layer scientists know best. Nearly three-quarters of its weight lies in two nonmetals: oxygen and silicon. Most of the rest consists of the six metals, aluminum, iron, calcium, sodium, potassium, and magnesium.

Within the crust most elements occur as minerals—natural substances that differ chemically and have distinct atomic structures. Earth's crust holds about 2,000 kinds of minerals. A few (including gold) occur as just one element; most comprise two or more elements chemically joined as compounds. Silicates (minerals containing silicon and oxygen) are the most abundant minerals in the crust and mantle, which make up four-fifths of our planet's volume. (See also pie chart, *below*.)

Crystal habits *(above)*
Three types of habit—crystal form determined by a crystal system but modified by factors such as temperature—are:
a Cubic
b Columnar
c Tabular

Crustal elements *(right)*
This pie chart shows the percentages of the most abundant elements in Earth's crust.
a Oxygen 46.6%
b Silicon 27.72%
c Aluminum 8.13%
d Iron 5.0%
e Calcium 3.63%
f Sodium 2.83%
g Potassium 2.59%
h Magnesium 2.09%
i Other elements 1.41%

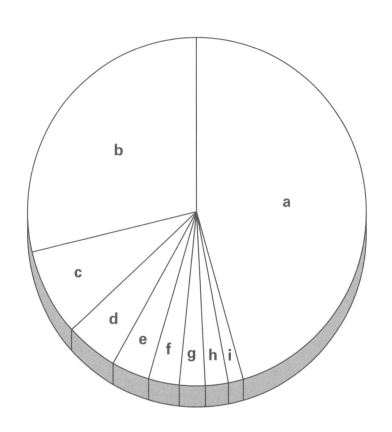

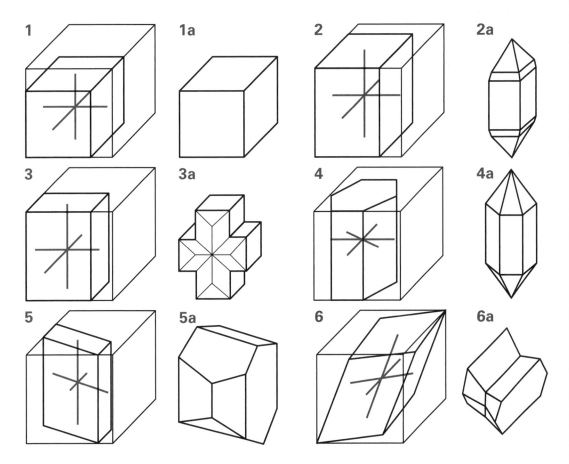

1
1a
2
2a
3
3a
4
4a
5
5a
6
6a

Most minerals formed from fluids that solidified—a process that arranged their atoms geometrically, producing crystals. Scientists identify six crystal systems based on axes—imaginary lines passing through the middle of a crystal. Each system yields crystals with distinctive symmetry. Within each system, each mineral crystal grows in a special shape or habit, though this can be modified by temperature, pressure, and impurities.

Traditional tests identify minerals by hardness, color, streak (color of the powdered mineral), luster, specific gravity, cleavage, fracture, form, tenacity (resistance to bending, breaking, and other forces), odor, taste, and feel. Sophisticated tests involve polarizing microscopes, X-rays, and spectral analysis.

Crystal systems
Systems vary with the number, lengths, and angles of axes. Internal axes are shown by colored lines.
1 Cubic
1a Example: halite
2 Tetragonal
2a Example: zircon
3 Orthorhombic
3a Example: staurolite
4 Hexagonal
4a Example: quartz
5 Monoclinic
5a Example: orthoclase
6 Triclinic
6a Example: albite

© DIAGRAM

EARTH'S BUILDING BLOCKS 2

A

B

C

Major rock types
Magnification reveals typical texture differences.
A Igneous rock: well-defined crystals
B Sedimentary rock: crystals worn by weathering and erosion
C Metamorphic rock: crystals aligned under stress.

Rocks are mixtures of minerals. Most rocks consist of interlocking grains or crystals stuck together by a natural cement. A few dozen minerals provide the main ingredients for the most common rocks. Here are some brief details about rock-forming minerals:

1 Silicates are the chief rock-forming minerals. Most feature a metal combined with silicon and oxygen. Examples: asbestos, mica, quartz, and feldspar.

2 Carbonates are the second most abundant group of minerals and include carbon, oxygen, and one or more metals. Examples: calcite, dolomite, and aragonite.

3 Sulfides are compounds of sulfur and one or more metals. Examples: galena and pyrite.

4 Oxides are compounds of oxygen and one or more metals. Examples: hematite and magnetite.

5 Halides are compounds of a halogen and a metal. Examples: Fluorite and halite (rock salt).

6 Hydroxides are compounds of hydrogen, oxygen, and usually a metal. Examples: limonite and brucite.

7 Sulfates are compounds of sulfur, oxygen, and a metal. Example: gypsum (the commonest sulfate).

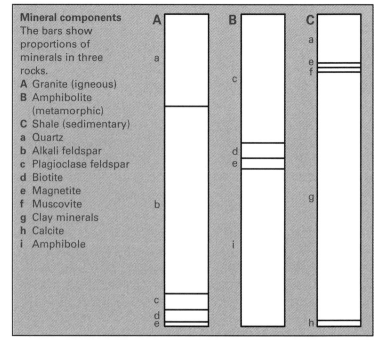

Mineral components
The bars show proportions of minerals in three rocks.
A Granite (igneous)
B Amphibolite (metamorphic)
C Shale (sedimentary)
a Quartz
b Alkali feldspar
c Plagioclase feldspar
d Biotite
e Magnetite
f Muscovite
g Clay minerals
h Calcite
i Amphibole

8 Phosphates are chemical compounds related to phosphoric acid. Examples: apatite and monazite.

9 Tungstates are salts of tungstic acid. Example: wolframite (a tungsten ore).

Rocks vary greatly in size, shape, and mineral proportions. But texture (influenced by origin) helps geologists identify three main groups: igneous, sedimentary, and metamorphic.

A Igneous rocks have interlocking crystals formed as molten rock cooled down. The smaller the crystals the faster the cooling.

B Sedimentary rocks mainly have rather rounded mineral grains joined by natural cements. Most derive from the deposited remains of older rocks.

C Metamorphic rocks have crystals with a tendency to banding or alignment. They are formed from older rocks recrystallized by heat and pressure.

Rock-forming minerals
Illustrations below depict nine rock-forming minerals representing nine groups. (Note: one mineral can occur in several forms.) Numbers match with those in the text.

1 Feldspar (a silicate)
2 Dolomite (a carbonate)
3 Pyrite (a sulfide)
4 Hematite (an oxide)
5 Fluorite (a fluoride)
6 Limonite (a hydroxide)
7 Gypsum (a sulfate)
8 Apatite (a phosphate)
9 Wolframite (a tungstate)

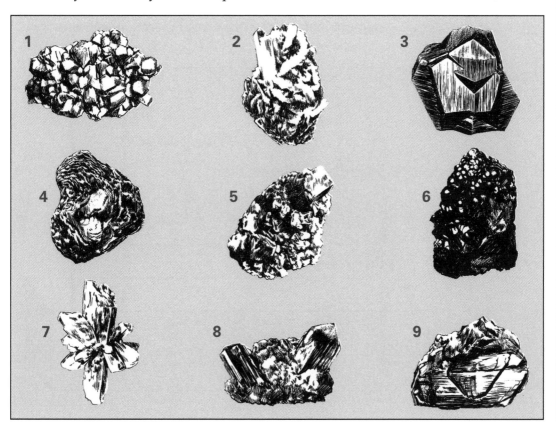

© DIAGRAM

27

1 ENERGY AND EARTH

Huge quantities of energy are always acting on Earth's surface and interior. Internal heat helps build and redeploy great sections of Earth's crust. External energy from the Sun and Moon keeps sea and air in motion, and sculpts the crustal surface.

The immense amount of heat trapped below Earth's crust has several origins, most dating from Earth's formation. Thus there is impact energy from colliding planetesimals, energy released by core formation, heat produced by outer layers pressing on the core, and radioactive energy from isotopes incorporated in the early Earth.

From Earth's core, convection currents convey heat through the mantle to the crust. More is added on the way by radioactive decay of crust and mantle minerals. Here some heat escapes in violent volcanic eruptions, hot springs, and earthquakes, but most simply leaks out quietly through continents and ocean floors—especially from the thin crust below some oceans.

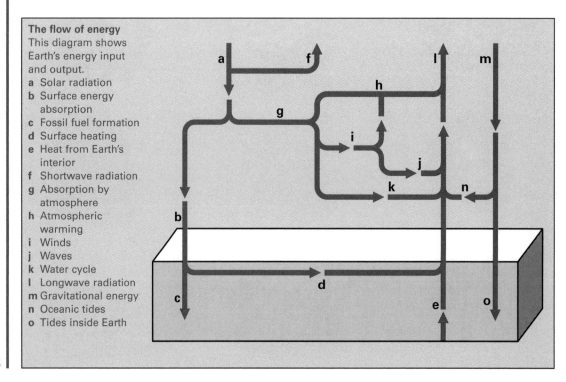

The flow of energy
This diagram shows Earth's energy input and output.
a Solar radiation
b Surface energy absorption
c Fossil fuel formation
d Surface heating
e Heat from Earth's interior
f Shortwave radiation
g Absorption by atmosphere
h Atmospheric warming
i Winds
j Waves
k Water cycle
l Longwave radiation
m Gravitational energy
n Oceanic tides
o Tides inside Earth

The surface of Earth's crust receives far more energy from above than below. Sunshine warms the atmosphere and crust. But sunshine warms the tropics more than the polar regions. This uneven heating creates belts of differing atmospheric pressure. Winds blow from high- to low-pressure areas. In turn, winds drive ocean waves and surface currents. Between them, winds and currents help to spread heat more evenly around the world. The Sun's heat also drives the water cycle. The resulting rain, rivers, glaciers, and ocean waves sculpt the surface of the land.

Besides gaining heat from above and below, the surface of Earth's crust loses heat through radiation into space. Loss roughly matches gain, so surface temperature has long remained about the same.

Lastly, the gravitational energy provided by the Moon and Sun produces our ocean tides and the tidal energy inside Earth itself.

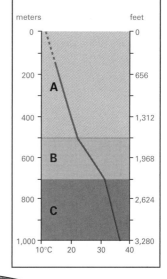

Temperature and depth
This graph shows temperature increases from 59°F (15°C) near the surface to 100°F (38°C) at a depth of 3,280 feet (1,000 m). Rate of increase varies with the rock, as in our examples.
A Limestone
B Shale
C Granite

Heat-flow map
The dark tone shows areas of highest heat flow (largely ocean ridges) from Earth's interior. Coolest areas include the old parts of continents and ocean basins.

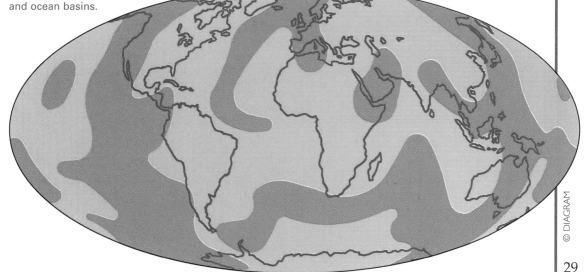

© DIAGRAM

EARTH AS A MAGNET

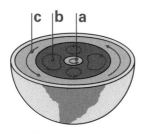

c b a

Inside Earth
Earth cut open at the equator shows the internal differences of rotation producing the magnetic field.
a Inner core rotation
b Eddies in the outer core
c Rotation of mantle

About 2,000 years ago the Chinese discovered that a freely turning lodestone spoon always ended pointing in the same direction. Such naturally magnetic lumps of iron oxide later led to the compass needle.

For centuries no one knew just why a magnetic compass worked. By AD 1600 experiments suggested that Earth itself exerted a magnetic force on compass needles. Observations proved that Earth indeed has a magnetic field the force of which aligns compass needles with north and south magnetic poles fairly near the geographic poles. (The field in fact extends far into space: this extension into space is called the magnetosphere.)

Just why Earth should be magnetic became clear only in the 1950s. Scientists now realize that Earth is less a huge bar magnet than a self-excited dynamo. Inside Earth, radioactive heat keeps streams of molten metal flowing through the outer core. This process generates electric currents that produce magnetic fields—much as an electric current flowing through a coil of copper wire creates a magnetic field around the

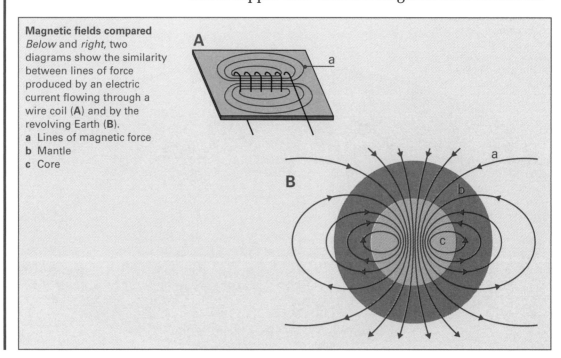

Magnetic fields compared
Below and *right*, two diagrams show the similarity between lines of force produced by an electric current flowing through a wire coil (**A**) and by the revolving Earth (**B**).
a Lines of magnetic force
b Mantle
c Core

wire. Earth's spin around its axis helps steer currents and create magnetic poles. Mighty eddies in the currents probably explain why magnetic poles slightly shift position from year to year. More puzzling are the hundreds of reversals of polarity occurring throughout Earth's history. Some rocks retain a record of Earth's polarity at the time those rocks were formed. Paleomagnetism—the study of Earth's magnetic field in prehistoric times—helps geologists date certain rocks. It also helps them understand past and present movements of vast slabs of crust—the major subject of our second chapter.

Polarity reversals
This column shows epochs of predominantly normal (**a**) and reversed (**b**) polarity for the last 4.2 million years.
A Brunhes (normal)
B Matuyama (reversed)
C Gauss (normal)
D Gilbert (reversed)

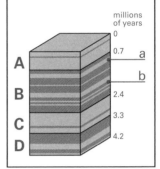

Earth's magnetosphere *(below)*
a Solar "wind" streamlining Earth's magnetic field
b Shock front where solar wind meets magnetic field
c Magnetopause: edge of the magnetic field
d Van Allen radiation belts
e Earth

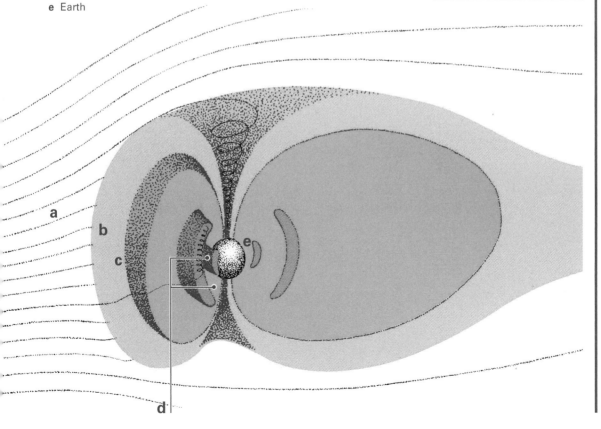

CHAPTER 2 THE RESTLESS CRUST

These pages explore the structure of Earth's crust—the thin, hard outer layer that we live on. We look at rocks that form the continents and ocean floor. We glimpse the slow processes that make and then destroy the ocean floor; build, move, and split the continents; and thrust up mountain chains. Pushing up and wearing down creates a grand recycling of rocks whose processes and products shape the land in ways described in later chapters.

A representation of ocean-bed topography in which ocean ridges and underwater contours can be clearly seen.

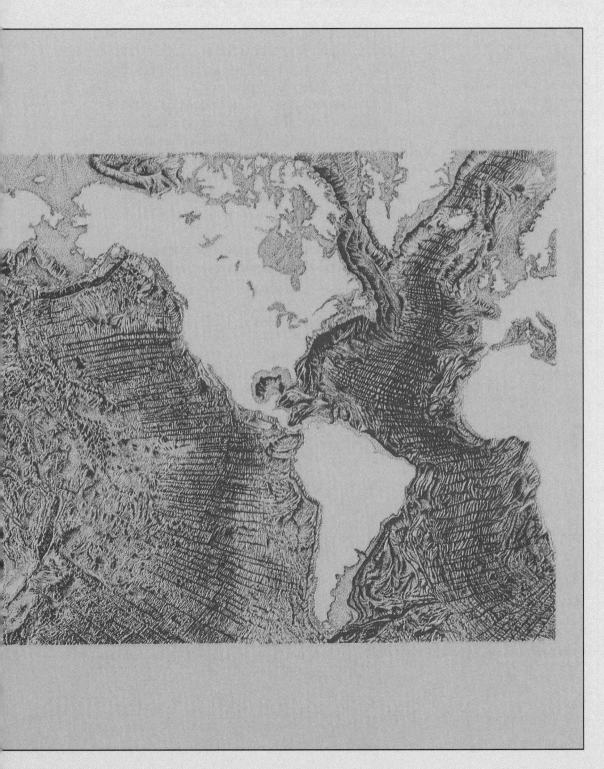

2 | EARTH'S CHANGING SURFACE

Plate tectonics in action
This cross section of the world suggests that lithosphere is made by currents rising in the mantle at constructive margins and lost at destructive margins where mantle currents sink.
a Continental crust
b Lithosphere
c Asthenosphere
d Lower mantle
e Core
f Constructive margin
g Destructive margin
h Lithospheric plate

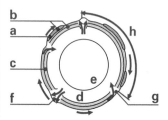

Tectonic plates
This map shows the major plates. Most plates are bounded by spreading ridges and collision zones or subduction zones (marked by oceanic trenches). Active boundaries give rise to earthquakes or volcanoes.

⊓⌐ Spreading ridges

— Collision zones

▲▲ Subduction zones

▨ Continental crust

•⦙• Volcanoes

⦙⦙ Earthquakes

a Eurasian plate
b African plate
c Antarctic plate
d Indo-Australian plate
e Pacific plate
f North American plate
g Nazca plate
h South American plate

Our planet's solid surface is a restless jigsaw of abutting, diverging, and colliding slabs called tectonic (or lithospheric) plates. How plates behave forms the subject known as plate tectonics.

Each plate involves a slab of oceanic crust, continental crust, or both, joined to a slab of rigid upper mantle. Collectively, these plates make up the lithosphere. This rides upon the asthenosphere, a dense, plastic layer of the mantle. Heat rising through this layer from Earth's core and lower mantle seemingly produces convection currents that shift the plates above.

Plate activities produce three main kinds of plate margins. Constructive margins are suboceanic

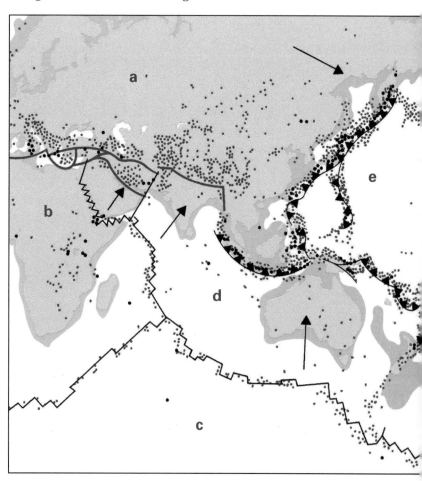

spreading ridges where new lithosphere is formed between two separating oceanic plates. Destructive margins are oceanic trenches where an oceanic plate dives down below a (less dense) continental plate. Conservative margins are where two plates slide past each other and lithosphere is neither made nor lost.

Geophysicists also talk about active margins (where colliding continental and oceanic plates spark off volcanic eruptions, earthquakes, and mountain building) and passive margins (tectonically quiet boundaries between continental and oceanic crust).

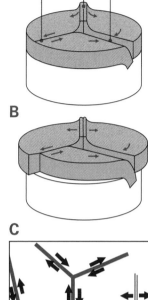

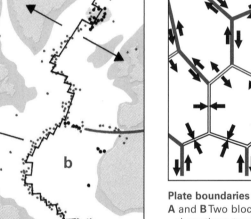

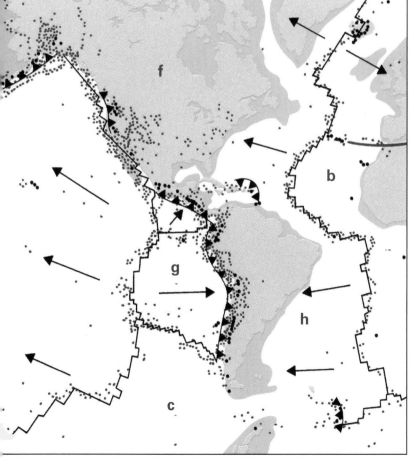

Plate boundaries
A and **B** Two block diagrams show three types of boundaries meeting as a triple junction between three tectonic plates.
A Plate positions now
B Plate positions later
a Conservative margins
b Constructive margins
c Destructive margin
C Bird's-eye view of possible boundary permutations where several plates interlock.

© DIAGRAM

2 | THE OCEAN FLOOR

Oceans and their seas hold 97 percent of all surface water, and cover some 71 percent of Earth to an average depth of 12,400 feet (3,800 m). Stripping off this watery sheath would reveal valleys, plateaus, peaks, and plains. Ten features of the ocean floor are explained below.

1 Continental shelf A continent's true but submerged and gently sloping rim; it descends to an average depth of 650 feet (200 m). Continental shelves occupy about 7.5 percent of the ocean floor.

2 Continental slope A relatively steep slope descending from the continental shelf. Such slopes occupy about 8.5 percent of the ocean floor.

3 Submarine canyon A deep cleft in the continental slope, cut by turbid river water flowing out to sea.

4 Continental rise A gentle slope below the continental slope.

The ocean floor
This cross section of an imaginary ocean floor depicts the 10 major features explained in the text on the opposite page. (The vertical scale is exaggerated for effect.)

1 Continental shelf
2 Continental slope
3 Submarine canyon
4 Continental rise
5 Submarine plateau/guyot
6 Abyssal plain
7 Seamount
8 Spreading ridge
9 Trench
10 Island arc

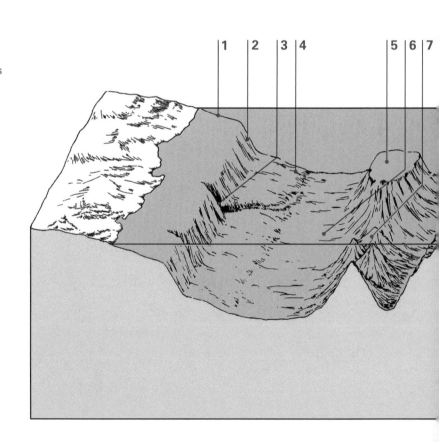

5 Submarine plateau A high seafloor tableland.

6 Abyssal plain A sediment-covered deep-sea plain about 11,500–18,000 feet (3,500–5,500 m) below sea level.

7 Seamount A submarine volcano 3,300 feet (1,000 m) or more above its surroundings. Guyots are flat-topped seamounts that were once volcanic islands.

8 Spreading ridge A submarine mountain chain generally 10,000 feet (3,000 m) above the abyssal plain. A huge system of such ridges extends more than 37,000 miles (60,000 km) through the oceans.

9 Trench A deep, steep-sided trough in an abyssal plain. At 35,840 feet (10,924 m) below sea level (deep enough to submerge Mount Everest), the Pacific Ocean's Marianas Trench is the deepest part of any ocean.

10 Island arc A curved row of volcanic islands, usually on the continental side of a trench.

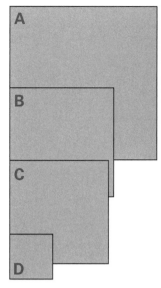

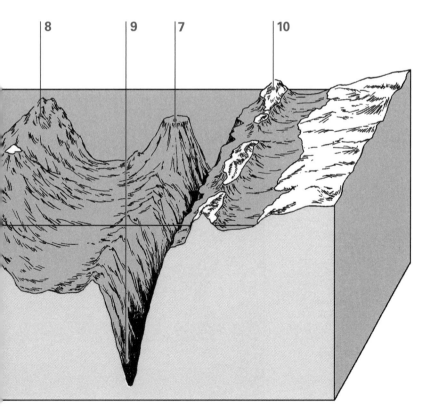

Ocean areas and depths
Diagrams contrast relative areas *(above)* and average depths *(below)* of Earth's four oceans, omitting their marginal seas.
A Pacific Ocean
B Atlantic Ocean
C Indian Ocean
D Arctic Ocean

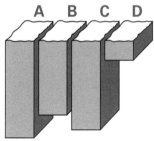

© DIAGRAM

2 | OCEANIC CRUST

The layered seabed
A block diagram shows one estimate of the average depths of the four layers of oceanic lithosphere riding on the mantle. (Layer 1 thins away from continents.) The numbers illustrate layers **1–4** in the text.

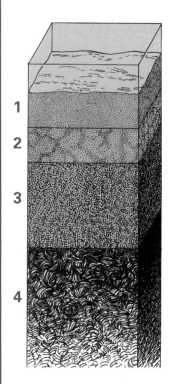

Dredging, boring, and seismic surveying suggest that oceanic crust is thinner, denser, and more simply made than continental crust. Oceanic crust is less than 6.2 miles (10 km) thick. Its rocks are richer than mantle rocks in aluminum and calcium, and their high silica and magnesium content earned oceanic crust the collective name of sima. Here are (simplified) details of oceanic crust's three layers and an associated fourth layer:

Layer 1 The top layer consists of sediments. Muds, sands, and other debris washed off continents lie up to half a mile (0.8 km) thick on continental shelves and nearby ocean floor. The open ocean's bed contains oozes (remains of dead microorganisms from the surface waters), clays, and (in places) nodules rich in

Ocean-floor sediments
A world map shows the distribution of marine sediments, which are largely the remains of land rocks and oozes formed from tiny dead marine organisms.

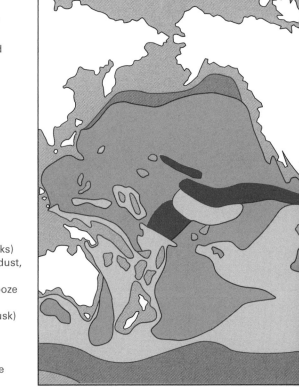

Terrigenous deposits (from eroded land rocks)
Red clay (from dust, etc.)
Foraminiferan ooze (calcareous)
Pteropod (mollusk) ooze
Diatom ooze (siliceous)
Radiolarian ooze (siliceous)

substances including manganese. No sediment occurs on spreading ridges.

Layer 2 is chiefly igneous rock, especially basalt, derived from the mantle and released at spreading ridges as rounded lumps of pillow lava. Scientists think the lower part of Layer 2 is seamed by sheeted dikes. Much of Layer 2 is 0.9–1.2 miles (1.5–2 km) thick.

Layer 3 is about 3 miles (5 km) thick and largely made of gabbro—a coarse-grained rock equivalent to the fine-grained basalt found in Layer 2.

Layer 4 is a rigid upper mantle layer connected to the bottom of the ocean crust. It may be largely made of the dense igneous rock peridotite, which chiefly consists of the mineral olivine.

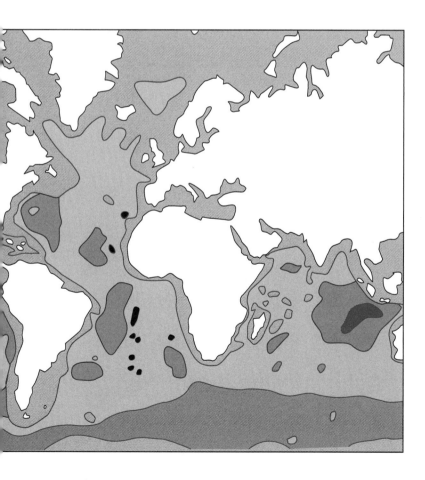

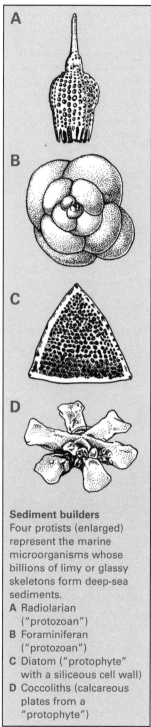

Sediment builders
Four protists (enlarged) represent the marine microorganisms whose billions of limy or glassy skeletons form deep-sea sediments.
A Radiolarian
   ("protozoan")
B Foraminiferan
   ("protozoan")
C Diatom ("protophyte" with a siliceous cell wall)
D Coccoliths (calcareous plates from a "protophyte")

© DIAGRAM

SEAFLOOR SPREADING

Spreading ridge evolving
Diagrams show how a spreading ridge grows.
1 Upwelling molten mantle melts and pushes up the oceanic crust above.
2 The bulging plate splits, and a central block subsides. Molten rock rises through cracks at the block's rims.
3 More molten rock plugs gaps left as tension pulls old crust apart.
4 As new crustal blocks subside, fresh cracks appear. The process is repeated.
a Asthenosphere (part melted).
b Peridotite upper mantle, formed from dense minerals.
c Gabbroic lower crust, from lighter minerals solidifying as lower ocean crust.
d Basalt and dolerite–fine-grained gabbroic rocks formed by fast cooling.

Scarcely any ocean floor is more than 200 million years old. Long ago a single mighty ocean incorporating the Pacific surrounded one landmass. The landmass developed splits that widened into basins. The Arctic, Atlantic, and Indian Oceans were created in this way.

Seafloor is always being made and destroyed by a process called seafloor spreading. Growth occurs at high heat-flow areas of oceanic crust where currents rising in the mantle hit the crust above. Seemingly this process helps tug apart vast chunks of crust, but the resulting gaps are continuously plugged by molten basalt and other rock originating in the mantle. Basalt sticking to the edges of such rifts formed the Mid-Atlantic Ridge and other vast underwater mountain chains called spreading ridges. Each widens by up to 10 inches (25 cm) a year.

Scientists believe that the detailed process goes as follows. First, molten rock wells up from deep down in the upper mantle region called the asthenosphere. The upwelling molten mass partly melts the rocks around it to make the oceanic crust. Gravity pulls the ridge flanks down and sideways. The resulting tension opens

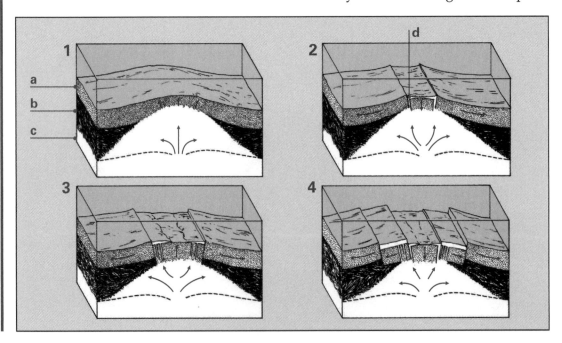

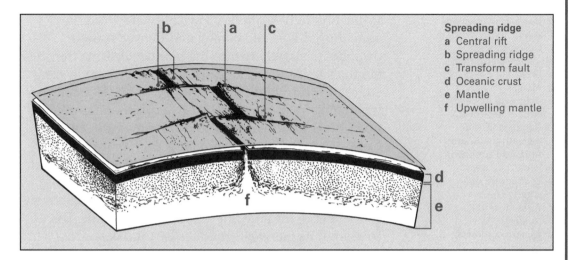

Spreading ridge
a Central rift
b Spreading ridge
c Transform fault
d Oceanic crust
e Mantle
f Upwelling mantle

two main cracks along the ridge. Between these cracks the ridge's middle sinks to form a central rift valley. Molten rock wells up through main and lesser cracks, then cools and hardens to become new ocean floor. Injections of fresh molten rock keep this spreading outwards from the central rift.

As upwelling continues, the rifting process is repeated. In time, rows of parallel ridges creep outwards, cooling, growing denser, and sinking down to form the ocean's abyssal plains.

Meanwhile, great cracks called transform faults cut across the central ridge at right angles, offsetting the short, straight sections.

Fossil magnetization pattern
a Central rift
b Normally magnetized basalt
c Reversely magnetized basalt. As upwelling basalt forms new seafloor it takes on the polarity of the magnetic field at that time. So a spreading seafloor records geomagnetic reversals occurring every few hundred thousand years.

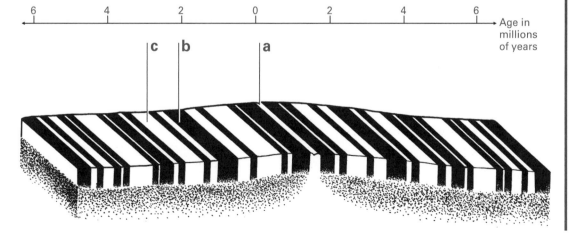

© DIAGRAM

HOW SEAFLOOR DISAPPEARS

While hot, light seafloor grows outward from the spreading ridges, old cooled and denser seafloor disappears elsewhere into the mantle. Oceanic trenches are the sites of these subduction zones, where leading edges of lithospheric plates plunge under less dense or less mobile plates and go below a continent or ocean floor. Most trenches lie around the rim of the Pacific Ocean; here the rocks originating from the spreading ridge that runs from Canada to south of New Zealand vanish.

Subducted oceanic crust injects a tongue of relatively cool material into the hot mantle rock beneath. The friction of its passage generates earthquakes.

As the subducted oceanic crust descends, its load of low-density sediments is largely scraped off and deformed. Meanwhile, 60 miles (100 km) down, the

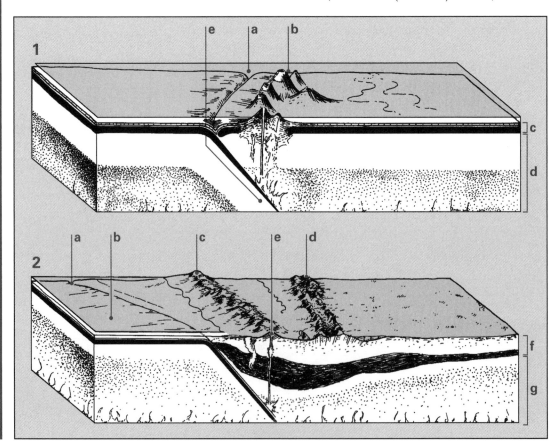

sinking lithosphere begins to melt; by 430 miles (700 km) down, it has completely broken up.

Less dense than the surrounding mantle, mantle melted by hot fluids from subducted lithosphere bobs up as it melts holes through the edge of the plate above the one subducted. Together with the scraped-off sediments, this process builds island arcs—rows of volcanic islands curved because they form upon Earth's curved surface. (Pressing a ping-pong ball with your thumb forms a similarly arc-shaped fold.) Examples of such arcs occur in the Aleutian, Japanese, Kurile, and Cycladic islands.

Oceanic crust subducted below a continental rim throws up volcanoes on the mainland. Volcanoes formed this way crown the Andes mountain chain of western South America.

Spreading and subduction
This diagram shows the complete conveyor belt sequence of ocean-floor production, transportation, and destruction.
a Spreading ridge
b Transform fault
c Subduction zone
d Oceanic crust
e Continental crust

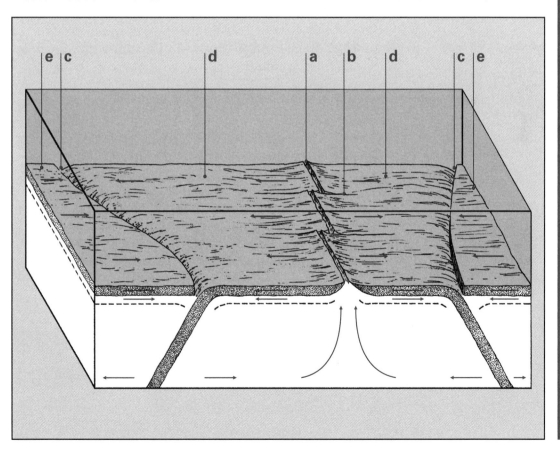

© DIAGRAM

2 | THE CONTINENTAL CRUST

The continents
Relative areas of continental lands above sea level:
a Australia
b Antarctica
c South America
d North America
e Africa
f Eurasia

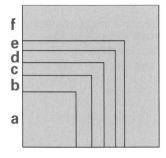

Inside a continent (*below*)
Section across an imaginary landmass:
a Continental shelf
b Young mobile belt (fold mountains with earthquakes and active volcanoes)
c Granitic/metamorphic rocks
d Granodioritic crust with intruded basic rocks
e Peridotite upper mantle
f Sedimentary basin
g Platform
h Shield
i Old mobile belt (old fold mountains without earthquakes or volcanoes).

Continents are the great landmasses above the level of the ocean basins. The six major masses are North America, South America, Eurasia, Africa, Australia, and Antarctica. With their submerged offshore continental shelves, these form 29 percent of Earth's surface and 0.3 percent of Earth as a whole.

Continents are thicker, less dense, and contain more complex rocks than ocean crust. Continental crust averages 20 miles (33 km) in thickness, but can be twice as deep below high mountains. The top 9 miles (15 km) or so consists of sedimentary, igneous, and metamorphic rocks rich in silicon and aluminum, hence the collective name *sial* often used for continental crust. The lower crust has denser igneous and metamorphic rocks. Continental lithospheric plates ride on the asthenosphere.

Continents average 2,950 feet (900 m) above sea level, but have wrinkles in the form of mountains, valleys, plains, and plateaus. According to geologists continents contain the following three main structural components:

1 Shields (or cratons) are stable slabs comprising outcrops of ancient masses of deformed crystalline rocks. Eroded shields plus overlying younger rocks are known as platforms. Shields and platforms form vast flat expanses, as in the plains of North America and Siberia, the Sahara Desert, the Congo Basin, and much of the Australian interior. Upraised parts of shields and platforms form high plateaus, especially in Africa and Asia.

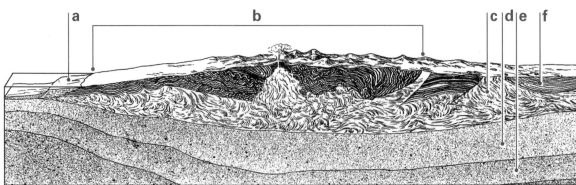

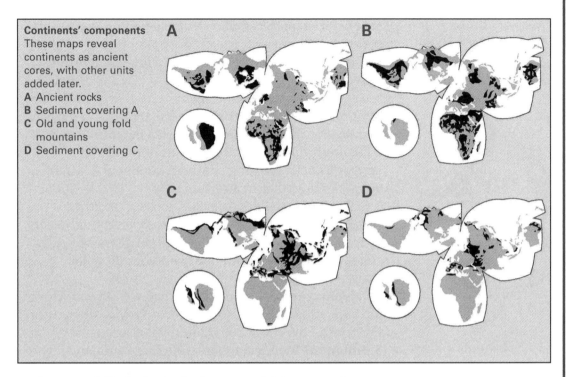

Continents' components
These maps reveal continents as ancient cores, with other units added later.
A Ancient rocks
B Sediment covering A
C Old and young fold mountains
D Sediment covering C

2 Linear mobile belts include young fold mountains like the Alps, Himalayas, Andes, and North American Cordillera. Beveled, mobile belts probably comprise the long rock structures seen around the rims of ancient shields.

3 Sedimentary basins are broad, deep depressions filled with sedimentary rocks formed in shallow seas that sometimes covered parts of ancient shields or their flanking mobile belts.

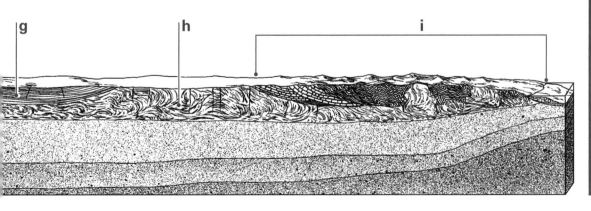

© DIAGRAM

CLUES TO CONTINENTAL DRIFT

1 Geographical clues
The Americas (**a**, **b**) fit into Europe (**c**) and Africa (**d**) if they are joined along their true rims, 6,600 feet (2,000 m) below the sea.

Scientists suspected that continents had moved around before the discovery of seafloor spreading showed how this shifting might have occurred. Here are five land-based clues to continental drift:

1 Geographical Some continents' coasts would almost interlock if they were rearranged like pieces of a jigsaw puzzle. For instance, South America fits into Africa.

2 Geological Old mountain zones of matching ages appear as belts crossing southern continents, if these are joined together in a certain way.

3 Climatic Glacial deposits and rocks scratched by stones in moving ice show that ice covered huge tracts of the southern continents 300 million years ago. This suggests that these landmasses once lay in polar regions.

4 Paleomagnetic Alignments of magnetized particles in old rock show that the southern continents all lay near the South Pole about 300 million years ago.

5 Biological Identical fossil land plants and land animals crop up in the southern continents now widely separated by the sea.

2 Geological clues
Aligning shields and rocks of three mountain-building phases suggest how southern lands may have once fitted together.

☐ Shields

▨ Early Paleozoic rocks

▨ Early Mesozoic rocks

▨ Late Mesozoic–Early Cenozoic

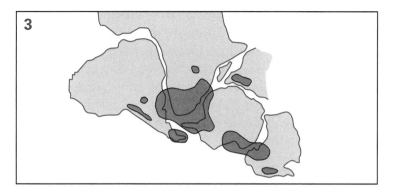

3 Climatic clues
About 320 million years ago South Polar ice sheets could have straddled southern lands in this way.

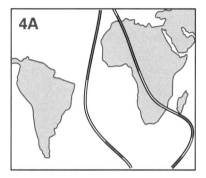

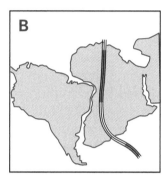

4 Paleomagnetic evidence
A The apparent wander paths of South America and Africa in relation to magnetic poles 400–200 million years ago.
B The paths coincided, if both continents were joined.

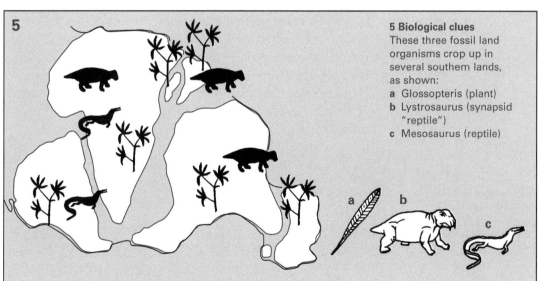

5 Biological clues
These three fossil land organisms crop up in several southern lands, as shown:
a Glossopteris (plant)
b Lystrosaurus (synapsid "reptile")
c Mesosaurus (reptile)

HOW CONTINENTS EVOLVE

A rifting continent *(below)*
A Continental crust intact
B Crust bulges above a rising
hot spot in the mantle
C Crust fractures
a Crust
b Mantle

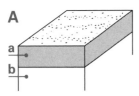

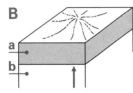

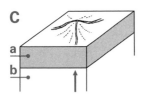

Close study of the rocks of continents reveals ancient cores with progressively younger rocks tacked on to their rims. Each core, or craton, originated as a microcontinent, possibly like this. Two converging, cooling, horizontal currents in the mantle tugged on a tract of thin, early crustal rock, then sank. This squashed and thickened that patch of crust. Its base bulged down and melted, releasing light material that punched up through the crust above. Such rock resorting could have formed the first small slabs of continental crust. Later, sea-floor spreading swept island arcs and sediments against microcontinents as mobile belts—belts of deformed and buckled rock. Accretion of this kind formed full-blown continents.

About five percent of today's continental crust had formed by 3.5 billion years ago, half by 2.5 billion years ago, most by 0.5 billion years ago. Once formed, continents are not immutable—they can be reworked, but not destroyed. Coalescing produced the supercontinent Pangaea about 300 million years ago. Rifting later broke it up.

Earth's crust splits open above "hot spots"—fixed plumes of molten rock rising in the mantle. A plume formed the volcanic Hawaiian Islands by punching through the thin oceanic Pacific plate passing over it. Plumes raise domes in the thick, rigid continental crust. A dome is liable to split in three as cracks grow outward from its top. Where three cracks widen, oceanic rock wells up into the spreading gaps. The continent is split apart, and a triple junction then

Triple junctions *(right)*
A A triple junction split South America from Africa.
B A triple junction splits Arabia from Africa. The Benue Trough (**a**) and Afar Depression (**b**) are both aulacogens.

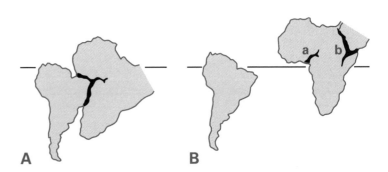

A a b c

B a b c

C d

separates three lithospheric plates. If spreading happens only in two cracks, two plates form. The third crack becomes an abandoned trough or rift. Nigeria's Benue Trough and Ethiopia's Afar Depression are two such so-called aulacogens.

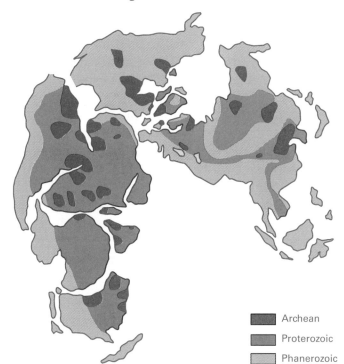

Archean

Proterozoic

Phanerozoic

Emerging microcontinents
(*above*)
Sections through Earth's curved surface suggest one way in which rock resorting of primal crust might have given rise to continental and oceanic crust.

A A greenstone island forms above a rising mantle current.

B A granitic craton forms above a sinking mantle current.

C A greenstone island belt, early oceanic crust, and embryonic craton result.

a Ocean
b Primal crust
c Mantle
d Oceanic crust

Accreting continents
This map shows continents growing in three phases: Archean (before 2,500 million years ago), Proterozoic (2,500–543 million years ago), and Phanerozoic (since 543 million years ago). Landmasses are pictured about 300 million years ago.

© DIAGRAM

49

MOUNTAIN BUILDING

Large regions of Earth consist of mountains. Most occur in rows called ranges. Parallel ranges and intervening plateaus form chains such as the Andes and North American Cordillera. Related mountain chains and ranges make up mountain systems, notably the Tethyan (Alpine-Himalayan) and Circum-Pacific systems.

Orogenesis, or mountain building, occurs along mobile belts—places where colliding lithospheric plates disrupt the continental crust. Such mountain-building belts are known as orogens and orogenic belts are belts of fold mountains—mountains created by crustal deformation and uplift. Geologically recent orogenic belts mostly rim continents. But ancient orogenic belts (the Ural Mountains for example) can occur deep inside a continent where lithospheric plates were welded together long ago.

Mountain building is a complex process. Deep troughs of accumulated offshore sediment, volcanic rocks, bits of oceanic crust, and scraps of foreign continents can all be swept against one continent and welded on as mountain ranges. Most of mountainous western North America consists of more than 50 suspect terranes—mighty slabs of alien rock that independently rotated and migrated north along the western edge of North America.

We illustrate three major mountain-building processes.

1 Oceanic plate subduction below another oceanic plate. This process created the Aleutian Islands and other mountainous island arcs.

2 Oceanic plate subduction beneath a continent. Involving island-arc collision, this process helped produce the Andes.

3 Double continent collision. The way in which the Alps and Himalayas formed.

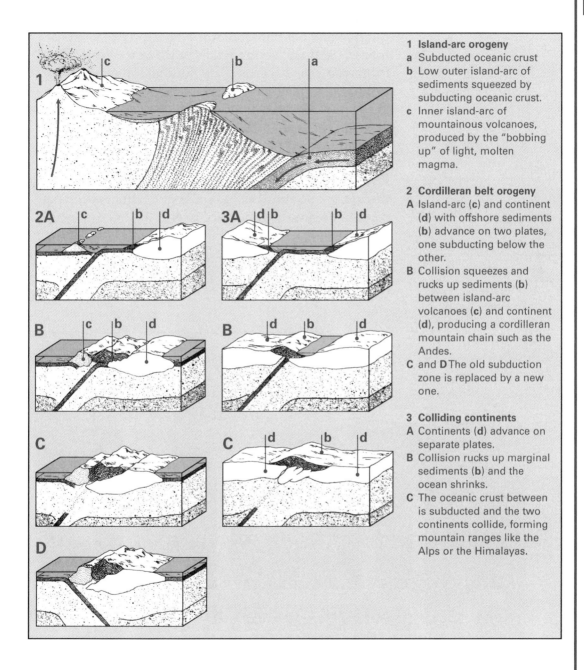

1 Island-arc orogeny
a Subducted oceanic crust
b Low outer island-arc of sediments squeezed by subducting oceanic crust.
c Inner island-arc of mountainous volcanoes, produced by the "bobbing up" of light, molten magma.

2 Cordilleran belt orogeny
A Island-arc (**c**) and continent (**d**) with offshore sediments (**b**) advance on two plates, one subducting below the other.
B Collision squeezes and rucks up sediments (**b**) between island-arc volcanoes (**c**) and continent (**d**), producing a cordilleran mountain chain such as the Andes.
C and **D** The old subduction zone is replaced by a new one.

3 Colliding continents
A Continents (**d**) advance on separate plates.
B Collision rucks up marginal sediments (**b**) and the ocean shrinks.
C The oceanic crust between is subducted and the two continents collide, forming mountain ranges like the Alps or the Himalayas.

ROCKS RECYCLED

Internal and external forces produce a rock cycle that builds, destroys, and remakes much of the rock that forms our planet's crust. The internal forces are produced by currents of rock that rise and spread out in the mantle, thereby moving lithospheric plates around. The main external forces are the weather, determined and generated by the energy in sunshine.

Weather wears down the rocks exposed above sea level. Rainfall creates rivers that transport rock debris to the sea, where it collects as sediment. Some sediments consolidate into sedimentary rock.

Materials

1 Molten rock in the mantle
2 Intrusive igneous rock
3 Extrusive (volcanic) igneous rock
4 Sediment
5 Sedimentary rock
6 Metamorphic rock

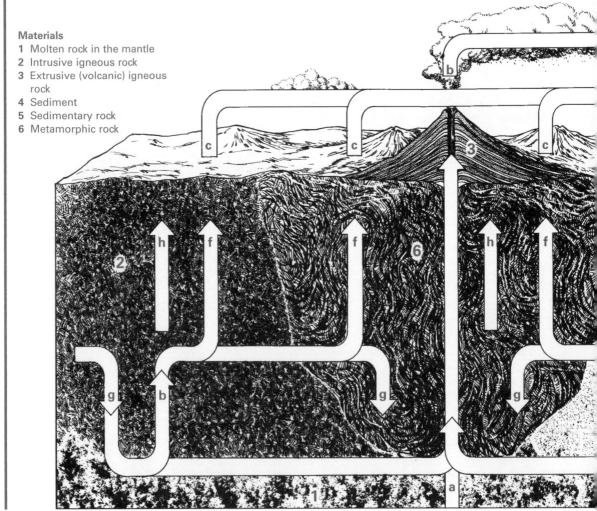

Colliding lithospheric plates thrust much of the sedimentary rock above the sea. Subducting lithospheric plates bear igneous and sedimentary rock down into the mantle, where heat and pressure turn it into metamorphic rock. Molten igneous rock rises through the cooler, denser rocks above, creating island arcs, injecting new material into the continental crust, and baking preexisting rocks.

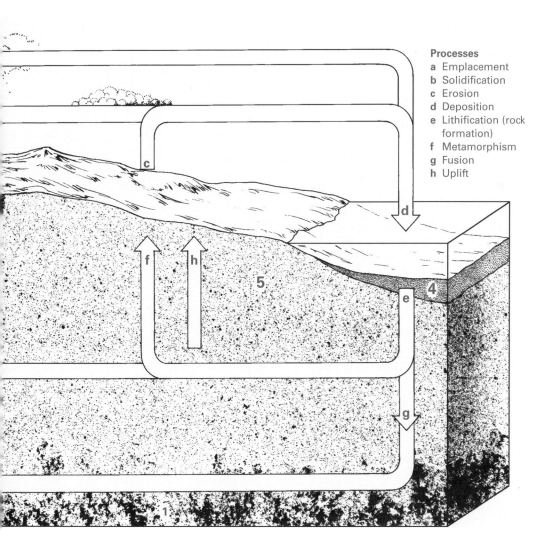

Processes
a Emplacement
b Solidification
c Erosion
d Deposition
e Lithification (rock formation)
f Metamorphism
g Fusion
h Uplift

© DIAGRAM

CHAPTER 3 FIERY ROCKS

Molten rock welling up from deep down in Earth's interior cools and hardens at or near the surface to create rocks such as lavas and granite. Igneous, or "fiery," rocks like these hold minerals that form the raw materials from which all crustal rocks derive. These pages describe major types and forms of igneous rocks, their ingredients, and the phenomena they yield—from mighty sheets and domes to tall volcanic cones, hot springs, and geysers. This chapter ends with a glimpse of fiery rocks from other planets.

An example of columnar basalt showing the characteristic hexagonal form. (Nineteenth-century engraving from *The National Encyclopaedia*)

ROCKS FROM MAGMA

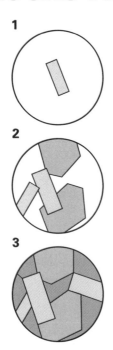

Igneous, or "fiery," rocks floor the world's oceans and form rock masses that rise from the roots of continents. Such rocks arise directly from the molten underground rock material—magma—that occurs where heat melts parts of Earth's upper mantle and lower crust. Most magma that has cooled and solidified escapes up through the crust from oceanic spreading ridges. Smaller quantities come from destructive plate boundaries and colliding continents.

Igneous rocks hold many minerals, chiefly silicates (silicon and oxygen usually combined with a base or metal). The major silicates are feldspars—silicates of aluminum combined with certain other elements, notably potassium (in alkali feldspars) and sodium and/or calcium (in plagioclase feldspars). Other

Crystals taking shape *(above)*
Minerals solidifying in sequence give igneous rocks an interlocking crystal texture
1 First mineral starts forming in molten magma
2 Second mineral forming
3 Last mineral to form fills any remaining space

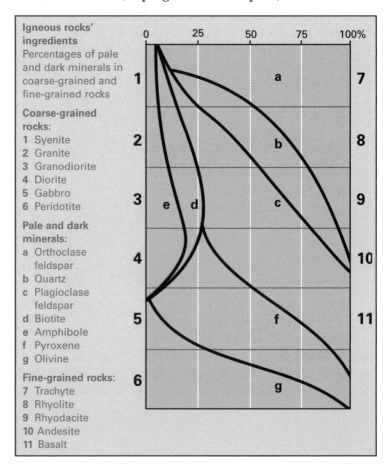

Igneous rocks' ingredients
Percentages of pale and dark minerals in coarse-grained and fine-grained rocks

Coarse-grained rocks:
1 Syenite
2 Granite
3 Granodiorite
4 Diorite
5 Gabbro
6 Peridotite

Pale and dark minerals:
a Orthoclase feldspar
b Quartz
c Plagioclase feldspar
d Biotite
e Amphibole
f Pyroxene
g Olivine

Fine-grained rocks:
7 Trachyte
8 Rhyolite
9 Rhyodacite
10 Andesite
11 Basalt

Magma production
Columns show annual magma output and loss in cubic kilometers from specific areas.
A Destructive plate boundaries (subduction zones)
B Constructive plate boundaries (spreading ridges)
C Within oceanic plates
D Within continental plates
E Plate material consumed at destructive plate boundaries

silicates include ferromagnesian minerals (rich in iron and magnesium); for instance, amphibole, biotite mica, olivine, and pyroxene are all ferromagnesian minerals. Quartz is the sole silicate comprising only silicon and oxygen.

Silica-rich igneous rocks are described as acid. In descending order of silica content, the other major groups are intermediate, basic (or mafic), and ultrabasic (or ultramafic).

Which type of rock evolves depends upon the type of parent magma and the processes it undergoes as it absorbs magma-melted rock, loses gas, and cools down. Different minerals "freeze out," or crystallize, and separate at different temperatures, so some magma yields rocks with layers of different minerals. But an igneous rock usually contains one set of minerals that solidified at roughly the same temperature. The rate of cooling influences crystal size; slow cooling gives large mineral crystals, and fast cooling yields fine-grained rocks.

The next pages describe igneous rocks formed in different conditions.

Solidifying minerals
Temperatures in °C at which seven minerals crystallize:
a Amphibole
b Quartz
c Olivine
d Mica
e Pyroxene
f Orthoclase feldspar
g Plagioclase feldspar

© DIAGRAM

FIERY ROCKS FORMED UNDERGROUND

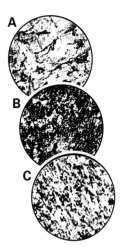

Three intrusive rocks
A Granite, an acid rock rich in quartz and feldspar
B Diorite, an intermediate rock, mainly plagioclase feldspar with darker minerals such as biotite or hornblende
C Gabbro, a basic rock, mainly plagioclase feldspar and pyroxene

Intrusive igneous rocks are produced where magma cools and hardens underground. Geologists place intrusive rocks in two categories: plutonic and hypabyssal.

Plutonic rocks include great masses formed deep in mountain-building zones; some develop from partial fusion of lower continental crust, and others come from magma rising from the mantle. Slow cooling produces big mineral crystals, usually coarse-textured rocks, including (acid) granite and granodiorite; (intermediate) syenite and diorite; (basic) gabbro; and (ultrabasic) peridotite. Granite—mainly made of quartz, feldspar, and mica—is the chief igneous rock of all continental crust. Erosion of overlying rock exposes granite masses like the domes above California's Yosemite Valley and the tors of Dartmoor in southwest England.

Hypabyssal rocks are relatively smaller masses, often strips or sheets. Such rocks cooled at a lesser depth and faster than plutonic rock, so they hold smaller crystals. Hypabyssal rocks include (acid) microgranite and microgranodiorite; (intermediate) microsyenite and microdiorite; and (basic) diabase (dolerite).

How granite forms
Light, molten blobs of granite rise from rocks melted deep down in the crust. Great blobs called plutons coalesce and cool as batholiths—immense rock masses in the cores of mountain ranges like the Sierra Nevada of California.

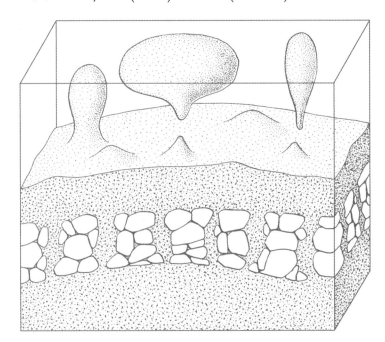

Intrusive rocks produce these features:

1 Batholith A huge deep-seated, dome-shaped intrusion, usually of acid igneous rock.

2 Stock Like a batholith but smaller, with an irregular surface area under 40 square miles (about 100 km²).

3 Boss A small circular-surfaced igneous intrusion less than 16 miles (26 km) across.

4 Dike A wall of usually basic igneous rock, such as diabase (dolerite), injected up through a vertical crack in preexisting rock.

5 Sill A sheet of usually basic igneous rock intruded horizontally between rock layers.

6 Laccolith A lens-shaped, usually acidic, igneous intrusion that domes overlying strata.

7 Lopolith A saucer-shaped intrusion between rock strata; can be up to hundreds of miles across.

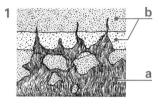
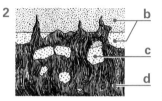

Stoping
1, 2: Granitic magma (**a**) rises by melting crustal rock (**b**). Xenoliths (**c**) are lumps of crust embedded in the granite (**d**).

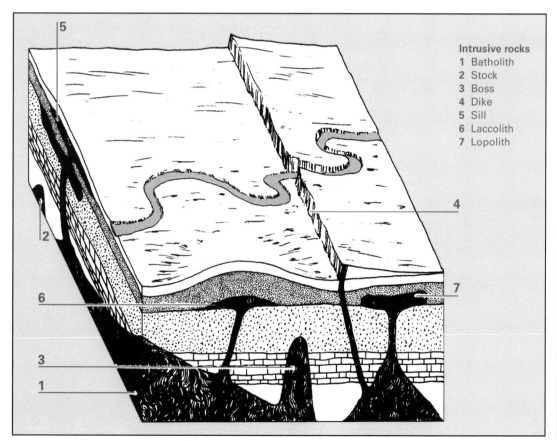

Intrusive rocks
1 Batholith
2 Stock
3 Boss
4 Dike
5 Sill
6 Laccolith
7 Lopolith

© DIAGRAM

VOLCANIC ROCKS

Basalt columns *(above)*
As basalt lava cools, it shrinks
and sometimes splits into
vertical columns. Famous
examples include Ireland's
Giant's Causeway and Staffa
in the Inner Hebrides.

Extrusive, or volcanic, igneous rocks occur chiefly at volcanic vents along the active margins of lithospheric plates. Here magma erupts as lava, which cools and hardens quickly on the surface as fine-grained or glassy rock.

Basic lavas are rich in metallic elements but poor in silica. They flow easily and erupt relatively gently. The best-known product is basalt, which accounts for more than 90 percent of all volcanic rock. This dark, fine-grained rock contains the minerals plagioclase feldspar, pyroxene, olivine, and magnetite. Basalt is formed by a partial melting of peridotite, the chief rock of the upper mantle. Basalt wells up from oceanic spreading ridges and builds new ocean floor; it also appears in rift valleys and rows of volcanoes like the Hawaiian Islands—perhaps products of a fixed plume of magma punching through the Pacific plate moving over it.

Acid (silica-rich) lavas appear at destructive plate margins. They probably comprise selected substances from basic lava of the upper mantle or reprocessed crust. Acid lavas are explosive and slow-flowing. They produce rocks such as dacite, rhyolite, and (black, glassy) obsidian.

Intermediate lavas contain plagioclase feldspar and amphibole, sometimes also alkali feldspar and quartz. They stem from partial melting of certain minerals in subducted oceanic crust. This process formed the lava andesite, named for the Andes Mountains. Found on

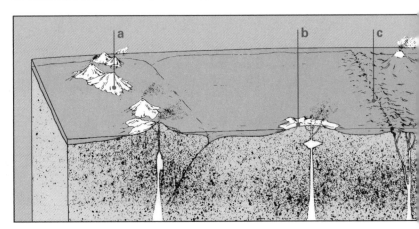

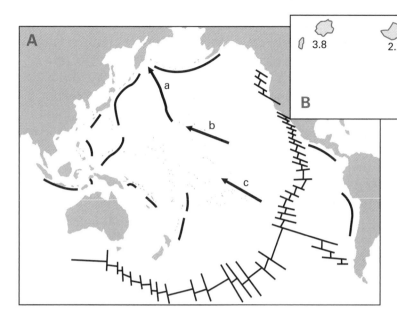

"Hot spot" volcanoes

A Arrows show rows of volcanoes perhaps punched through the moving Pacific plate by fixed "hot spots" — magma plumes. (The same hot spot arguably produced **a-b**). The newest volcanoes are those farthest from the arrow tips.

a Emperor Seamounts
b Hawaiian Islands
c Pitcairn-Tuamotu group

⊢⊢ Spreading ridges

╱ Ocean trenches

B Hawaiian Islands. The numbers are ages of the youngest volcanic rocks in millions of years.

the landward side of oceanic trenches, andesite builds island arc volcanoes and tacks new land onto the rims of continents.

More than 850 known volcanoes have erupted in the last 2,000 years. Those emitting continuously or from time to time are active; many form what is referred to as the "ring of fire" around the Pacific Ocean. Volcanoes not erupting in recent times are known as dormant. Long-inactive volcanoes are said to be extinct; some occur where colliding plates fused together many millions of years ago.

Where volcanoes erupt
a Island arc
b Shield volcano
c Spreading ridge
d Cordilleran mountains
e Lava plateau
f Rift valley

© DIAGRAM

ANATOMY OF A VOLCANO

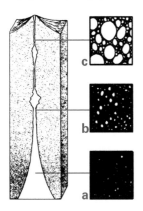

Gas and magma *(above)*
Inside a volcano as magma rises its pressure falls. Dissolved gases escape and form expanding bubbles (a, b, and c). These force magma out of the volcano.

How Parícutin grew
This diagram shows the rapid growth of Parícutin, Mexico, in height and distance from the center of the cone.
a After 24 hours
b After one week
c After one year
d Parícutin village church

Volcanoes take two main forms. Fissure, or linear, volcanoes chiefly emit basic lava from a crack in Earth's crust. Central volcanoes yield lava, ash, and/or other products from a single hole. These products build a shield- or cone-shaped mound—the typical volcano shape. Central volcanoes can grow high and fast. In western Mexico in 1943, Parícutin grew 490 feet (150 m) high in a week and reached 1,500 feet (450 m) in a year. In western Argentina, extinct Aconcagua towers 22,834 feet (6,960 m) above sea level; this is the highest mountain in the Western Hemisphere.

A cross section through an active central volcano would reveal these features: miles below the surface lies the magma chamber, a reservoir of gas-rich molten rock under pressure. This pressurized magma may "balloon" outward against the surrounding solid rock until it can relieve the pressure by escaping through a weakness in the crust above. From the chamber, magma then rises through a central conduit. As magma rises, the pressure on it is reduced, and its dissolved gases are freed as expanding bubbles. Finally the force of gases blasts open a circular vent on Earth's surface. From this outlet ash, cinders, and flows of lava build the main volcano shield or cone. Vent explosions shape its top as an inverted cone or crater. Meanwhile, side vents on the flanks of the volcano release ash or lava that may build subsidiary cones.

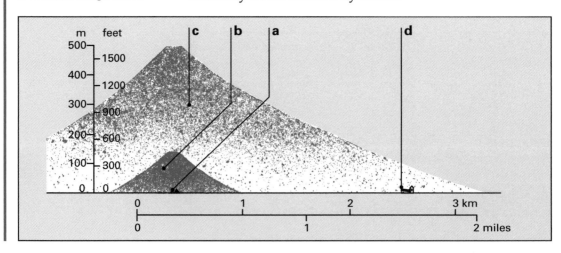

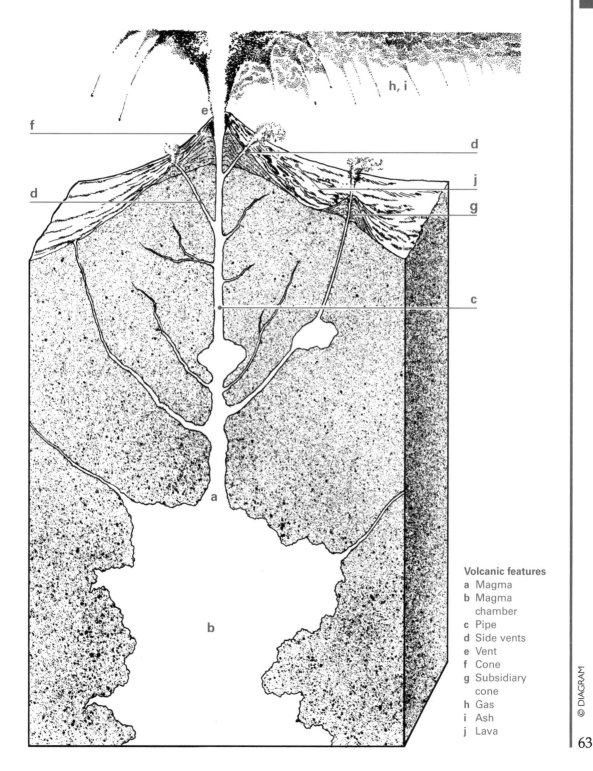

Volcanic features
a Magma
b Magma
  chamber
c Pipe
d Side vents
e Vent
f Cone
g Subsidiary
  cone
h Gas
i Ash
j Lava

© DIAGRAM

VOLCANIC LANDFORMS

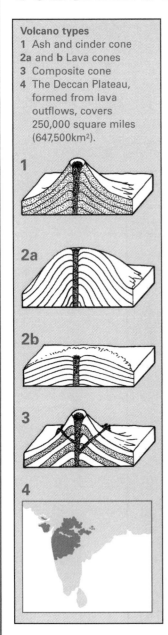

Volcano types
1 Ash and cinder cone
2a and b Lava cones
3 Composite cone
4 The Deccan Plateau, formed from lava outflows, covers 250,000 square miles (647,500km²).

Surface features of volcanic origin range from towering peaks and vast sheets of lava to craters small and low enough to jump across. Features vary depending on the type of eruption, material erupted, and effects of erosion. There are four major types of volcanic landform:

1 Ash and cinder cones, or explosion cones, occur where explosive eruption ejects solid fragments from a central crater and/or subsidiary craters. The resulting concave cone is seldom higher than 1,000 feet (300 m). Idaho's Craters of the Moon National Monument has many such examples.

2 Lava cones usually form from slowly upwelling lava. They come in two main types:

a Steep-sided volcanoes, like France's Puy de Dôme and Lassen Peak in California, grew from sticky acid lava that soon hardened. When squeezed out like toothpaste, very viscous lava builds spines like Wyoming's Devils Tower.

b Shield volcanoes (gently sloping domes) form from runny lava that flowed far before it hardened. From a seabed base 500 miles (800 km) across, Hawaii's Mauna Loa gradually rises 30,000 feet (9,144 m) to the broad crater at its summit.

3 Composite cones, or strato-volcanoes, have concave, cone-shaped sides that feature alternating ash and lava layers. Composite cones account for most of the highest volcanoes. Mount Fuji, Mount Rainier, and Vesuvius are three well-known examples. If solid lava plugs the main pipe to the crater, pent-up gases may blast the top off. If the magma chamber empties, the summit may collapse. The product of either occurrence is a vast shallow cavity called a caldera. Calderas include Crater Lake in Oregon, Tanzania's Ngorongoro Crater, and Japan's Aso, whose 71 mile (112 km) circumference makes this the largest caldera in the world.

4 Plateau basalts, or lava plains, occur where fissures leaked successive flows of basic lava that have blanketed huge areas in basalt. Basalt up to 7,000 feet (2,100 m) thick covers 250,000 square miles

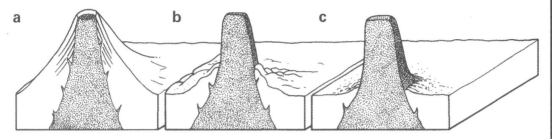

(650,000 km²) in India's Deccan Plateau. Other outflows form the U.S. Columbia River Plateau, South America's Paraná Plateau, the Abyssinian Plateau, and Northern Ireland's Antrim Plateau.

Volcanic plug
a Upwelling lava fills the original volcano's central pipe.
b Erosion attacks the soft outer slopes.
c Only the resistant lava plug remains.

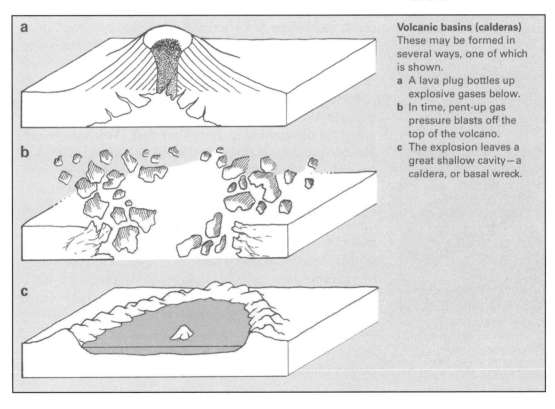

Volcanic basins (calderas)
These may be formed in several ways, one of which is shown.
a A lava plug bottles up explosive gases below.
b In time, pent-up gas pressure blasts off the top of the volcano.
c The explosion leaves a great shallow cavity—a caldera, or basal wreck.

© DIAGRAM

VOLCANIC PRODUCTS

A

B

How lavas behave *(above)*
Lavas vary in viscosity—
resistance to internal flow.
A Low-viscosity lava flows
readily, like water.
B High-viscosity lava flows
sluggishly, like molasses.

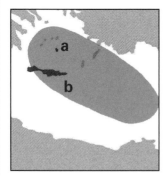

Volcanic fallout *(above)*
About 1628 BC, as redated in
1988, a huge volcano
exploded on the Aegean
island Thíra, also known as
Santorini (**a**). The ash-fall
(shown here tinted) covered
much of Crete (**b**) and might
have helped destroy Crete's
great Minoan civilization.

Volcanoes produce gases, liquids, and solids.

Volcanic gases include steam, hydrogen, and sulfur as well as carbon dioxide. Steam condensing in air forms clouds that shed heavy rain. Interacting gases intensify the heat in erupting lavas, and explosive eruptions may yield nuées ardentes—burning clouds of gas with scraps of glowing lava.

Liquid lava is the main volcanic product. Acid, sticky lava cools and hardens before flowing far. It may block a vent, causing magmatic pressure build-up, relieved by an explosion. Basic, fluid lava flows far and quietly before hardening, especially if it is rich in gas.

Varying conditions produce the following lava forms: **Aa** features chunky blocks formed where gas spurted from sluggish molten rock capped by cooling crust. **Pahoehoe** has a wrinkly skin made by molten lava flowing fast below it. **Pillow lava** resembles pillows and piles up where fast-cooling lava erupts underwater.

Solid products of explosive outbursts are called **pyroclasts**. These can be fresh material or ejected scraps of old hard lava and other rock. **Volcanic bombs** include pancake-flat scoria shaped on impact with the ground, and spindle bombs twisted by whizzing through the air. Acid lava full of gas-formed cavities produces **pumice**, a volcanic rock light enough to float on water. **Ignimbrite** contains naturally welded glassy fragments. Hurled-out cinder fragments are called **lapilli**. Some volcanoes also spew vast clouds of **dust** or **volcanic ash**—tiny lava particles. Ash mixed with heavy rain produces mudflows like the one that buried the Roman town Herculaneum when Mt. Vesuvius erupted in AD 79. Immense explosions can smother land for miles in ash and hurl vast quantities of dust into the higher atmosphere, cooling climates on a global scale and adding layers to deep ocean sediments.

Violent eruptions destroy towns and farms, but volcanic ash also provides rich soil for crops.

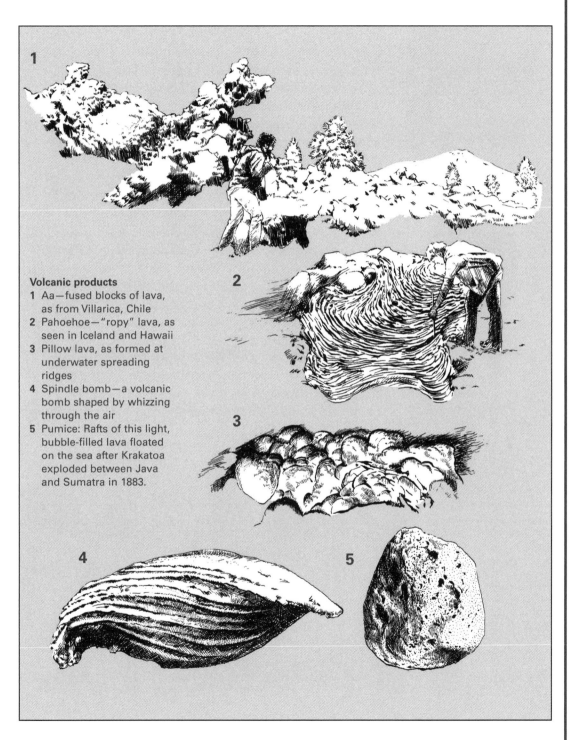

Volcanic products

1 Aa—fused blocks of lava, as from Villarica, Chile
2 Pahoehoe—"ropy" lava, as seen in Iceland and Hawaii
3 Pillow lava, as formed at underwater spreading ridges
4 Spindle bomb—a volcanic bomb shaped by whizzing through the air
5 Pumice: Rafts of this light, bubble-filled lava floated on the sea after Krakatoa exploded between Java and Sumatra in 1883.

HOT WATER, GAS, AND MUD

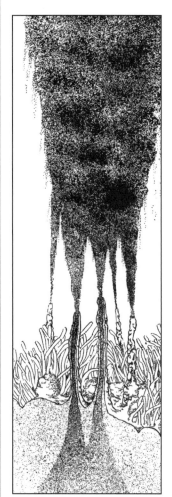

Smokers *(above)*
Jets of sulfide-rich water heated to 650°F (350°C) escape from mineral chimneys in the ocean floor—a phenomenon of spreading ridges.

Sinter terraces *(right)*
These steps consist of minerals deposited by mineral-rich water escaping from the ground as hot springs.

Hot water, gas, and mud squirt or dribble from vents in the ground heated by volcanoes mostly near extinction. Such features are plentiful in parts of Italy, Iceland, New Zealand, and the United States. Here are brief definitions of these forms:

Hot spring Spring water heated by hot rocks underground. Hot springs shed dissolved minerals that produce sinters (crusts) of (calcium carbonate) travertine or (quartz) geyserite. Famous hot springs occur in Iceland, New Zealand's North Island, and Yellowstone National Park.

Smoker Submarine hot spring at an oceanic spreading ridge; the best known is the Galápagos Rise. Emitted sulfides build chimneys that belch black, smoky clouds.

Geyser Periodic fountain of steam and hot water forced up from a vent by water superheated in a pipe or cave deep down. Famous geysers occur in Iceland and Yellowstone National Park.

Mud volcano Low mud cone deposited by mud-rich water escaping from a vent. Iceland, New Zealand's North Island, and Sicily have mud volcanoes.

Fumarole Small vent emitting jets of steam, as at Mt.

Etna, Sicily, and in Alaska's Valley of Ten Thousand Smokes.

Solfatara Volcanic vent emitting steam and sulfurous gas; named after one near Naples, Italy.

Mofette Small vent emitting gases including carbon dioxide. Examples occur in France (Auvergne), Italy, and Java.

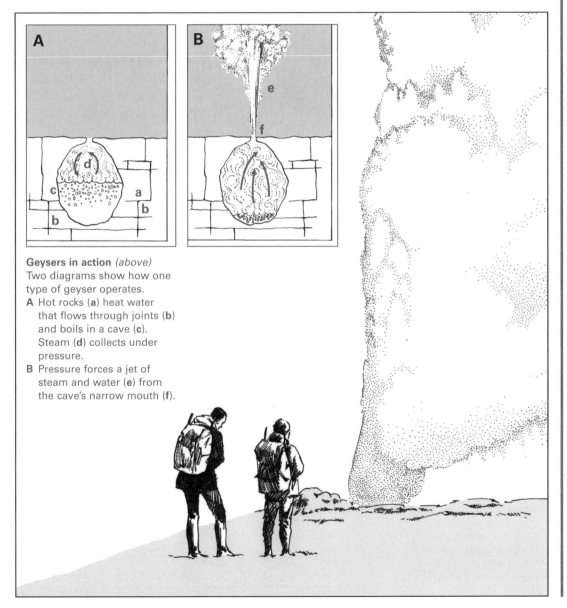

Geysers in action *(above)*
Two diagrams show how one type of geyser operates.

A Hot rocks (**a**) heat water that flows through joints (**b**) and boils in a cave (**c**). Steam (**d**) collects under pressure.

B Pressure forces a jet of steam and water (**e**) from the cave's narrow mouth (**f**).

3 FIERY ROCKS OF OTHER WORLDS

Beyond Earth, space probes have revealed an igneous rocky crust on the solid-surfaced planets Mercury, Venus, and Mars, and on some moons. Certain worlds possess volcanoes, though most of their craters were dug out by meteorites.

The landscape of much-cratered Mercury includes smooth plains, perhaps old basalt lava flows.

The surface of Venus *(above)* As seen from Russian space probe *Venera 14*, the Venusian surface revealed layered, weathered basaltic rocks similar to those on Earth.

Basalt lava probably built the plains that largely cover Venus. Radar shows vast twin cones called Rhea Mons and Theia Mons. Perhaps the largest (shield) volcanoes in the solar system, they may contribute to the sulfuric-acid clouds that cloak this planet. Basalt lava plains sprawl over much of northern Mars. Mars also has four mighty, very old volcanoes. Olympus Mons, the highest, towers at least 14 miles (23 km) above the Martian plains. Only Venus's giant volcanoes may be larger.

The Moon's surface shows two major types of rock. Its pale highlands consist mainly of anorthosite (found on Earth only in old parts of continents) and related rocks rich in plagioclase feldspar. The Moon's dark "seas," or *maria*, are ancient basalt lava flows. These welled up, filling basins gouged out by the impact of asteroids or mini-moons. The Moon is now volcanically dead. But meteorite impacts melting surface rocks created breccias consisting of sharp stones in a glassy matrix.

The solar system's most volcanically active world is Io—a moon of Jupiter the same size as our Moon. Space probe images show hundreds of volcanic craters and some immense volcanoes. Hot rocks heat sulfur in pockets underground. The molten sulfur rises, melts sulfur dioxide in pipes above, then squirts it out like

Mars' giant mountain *(below)* This diagram contrasts the relative heights and sizes of three volcanoes:

a Olympus Mons, Mars
b Everest (Earth's highest peak above sea level)
c Mauna Loa, Hawaii. (Lies mostly below the Pacific Ocean)

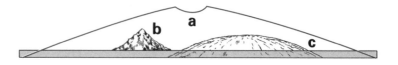

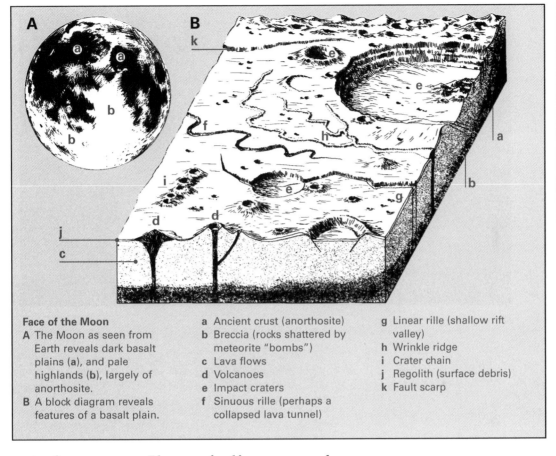

Face of the Moon

A The Moon as seen from
Earth reveals dark basalt
plains (**a**), and pale
highlands (**b**), largely of
anorthosite.

B A block diagram reveals
features of a basalt plain.

a Ancient crust (anorthosite)
b Breccia (rocks shattered by
meteorite "bombs")
c Lava flows
d Volcanoes
e Impact craters
f Sinuous rille (perhaps a
collapsed lava tunnel)

g Linear rille (shallow rift
valley)
h Wrinkle ridge
i Crater chain
j Regolith (surface debris)
k Fault scarp

water from a geyser. Plumes of sulfur compounds
reach 200 miles (300 km) above the surface. Then they
splash back, coloring the surface yellow, orange, black,
and white like some colossal pizza.

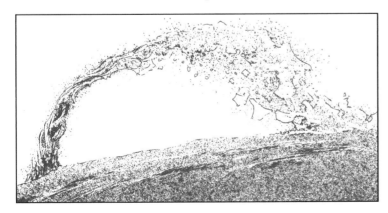

Sulfur fountains
Molten sulfur compounds
spurt high above the ever-
active surface of a strange
moon, Io.

CHAPTER 4 ROCKS FROM SCRAPS

When weather breaks up igneous or other surface rocks, wind and rivers bear away their broken scraps. Most settle on the seabeds fringing continents. Heavy pebbles get washed just offshore; sand farther out; and light, fine particles of silt and clay farther still. Pressure and natural cements convert these layered sediments to conglomerate, sandstone, and shale rock types. Meanwhile compacted remains of swamp plants and shallow-water animals form coals and limestone. Sedimentary rocks form only five percent of Earth's crust, but they cover three-quarters of its land.

1 A coquina quarry on Anastasia Island, Florida. Coquina, a white limestone formed of broken shells and corals is often used as a building material.
2 The white cliffs along the southeast coast of England were formed from chalk, a soft limestone chiefly composed of the shells of coccoliths. (Engraving of coquina quarry from *Picturesque America* 1894. Pictures of chalk cliffs from the Mansell Collection)

4

ROCKS FROM SEDIMENTS

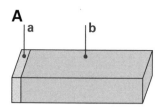

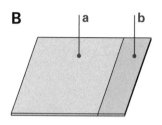

Rock production *(above)*
A Relative volumes of (**a**) sedimentary and changed sedimentary rocks and (**b**) igneous and changed igneous rocks in Earth's crust contrast with:
B The same rocks' exposed areas

Sediment or sedimentary rock covers most of the ocean floor and three-quarters of the land. On land this skin is usually a few miles thick, but layers up to 19 miles (30 km) thick collect in offshore basins. Most sedimentary rock comes from scraps of older (igneous or other) rocks eroded from the land, carried into lakes or seas by rivers, deposited, and then consolidated into a solid mass. When parent rock breaks up its minerals behave in different ways. Some of the silicates (the main mineral ingredients of igneous rocks) dissolve; others—quartz, for one—endure; and weathering creates new minerals, especially the clays that bulk large in most sedimentary rock. Besides the clastic sedimentary rocks—rocks made from fragments—others come from chemical precipitates or from the durable remains of living things.

Processes converting sediment to rock are known as diagenesis. Two main processes occur. As sediments pile up, their pressure squeezes water from the sediments below and packs the particles together. Then, too, some minerals between the grains cement a mass of sediment together.

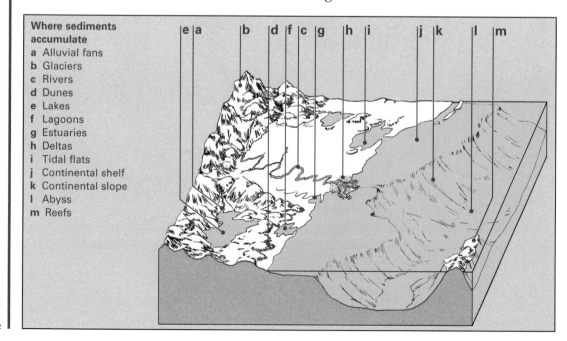

Where sediments accumulate
a Alluvial fans
b Glaciers
c Rivers
d Dunes
e Lakes
f Lagoons
g Estuaries
h Deltas
i Tidal flats
j Continental shelf
k Continental slope
l Abyss
m Reefs

Cross-bedding in Utah
A human figure shows the scale of cross-bedded sandstone sediments in Zion National Park.

Two types of bedding
A Graded bedding: large particles had time to settle before small particles
B Cross-bedding in sands laid down by migrating dunes or ripples

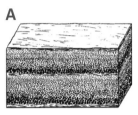

A

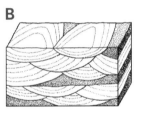

B

Changes converting sediment to rock leave traces in the finished product. Transportation of eroded sediments abrades and rounds their particles, sorts these by density or size, "rots" unstable minerals, and concentrates resistant minerals, including diamonds and gold.

Deposition lays down sediments in broadly horizontal sheets called beds or strata, each separated from the next in the pile by a division called a bedding plane. Beds with ripple marks reveal ancient currents. Graded bedding (beds with grain size graded vertically) may hint at turbidity currents—sediment-rich water sliding soupily down a continental slope. Cross bedding (sands laid down at an angle between two bedding planes) shows features such as old dunes and sandbars.

Diagenetic processes *(below)*
1 Before compaction
2 After compaction
3 Before cementation
4 After cementation

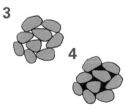

☐ Clay particles
▨ Mineral particles
■ Natural cement

© DIAGRAM

75

4 | ROCKS FROM FRAGMENTS 1

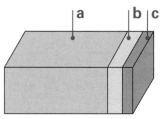

Sedimentary rocks *(above)*
Sedimentary rocks can be
divided into the following:
a Shale: 81%
b Sandstone: 11%
c Limestone: 8%

Most sedimentary rocks form from particles eroded from the rocks on land. Their main ingredients are clasts—rock fragments—of quartz, feldspar, and clay minerals. These fragments range in size from microscopic grains to boulders.

More than 90 percent of all sedimentary rock contains particles no bigger than a sand grain. Many geologists classify such partides by size into two main groups. The (fine-grained) lutites with grains less than 0.06 mm diameter produce mudstone, siltstone, and shale. The (medium-grained) arenites or sandstones with grains of 0.06–2 mm give arkose, graywacke, and orthoquartzite. Six fine- and medium-grained rocks are:

Mudstone Soft rock made of clay minerals of less than 0.004 mm diameter.

Siltstone Rock formed of particles 0.004–0.06 mm in diameter.

Shale Mudstone, siltstone, or similar fine-grained rock of silt and clay split easily along their bedding planes. Shale accounts for more than 80 percent of all sedimentary rock.

Orthoquartzite A "clean," or pure, arenite mainly made of quartz after other substances have been

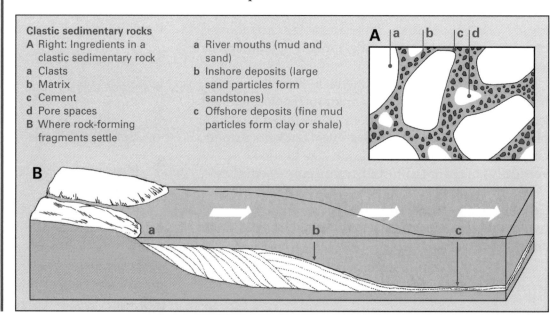

Clastic sedimentary rocks
A Right: Ingredients in a clastic sedimentary rock
a Clasts
b Matrix
c Cement
d Pore spaces
B Where rock-forming fragments settle

a River mouths (mud and sand)
b Inshore deposits (large sand particles form sandstones)
c Offshore deposits (fine mud particles form clay or shale)

removed. (Arenites account for more than ten percent of all sedimentary rock.)

Arkose An arenite rich in feldspar derived from gneiss or granite.

Graywacke A muddy, often grayish sandstone with mixed-size particles including quartz, clay minerals, and others.

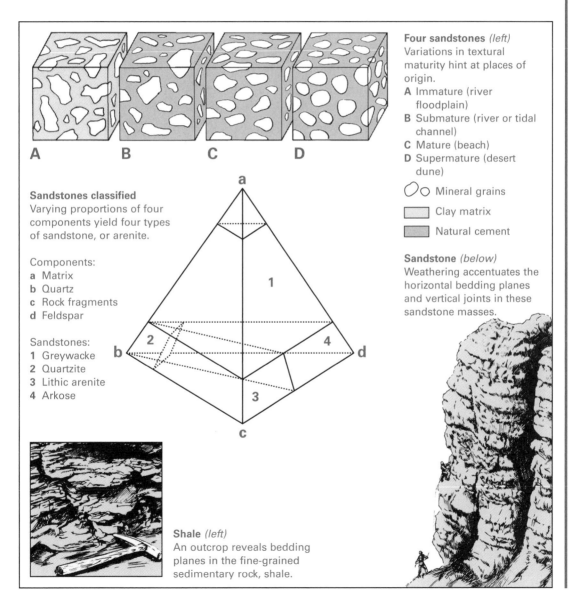

Four sandstones *(left)*
Variations in textural maturity hint at places of origin.
A Immature (river floodplain)
B Submature (river or tidal channel)
C Mature (beach)
D Supermature (desert dune)

⬭⭕ Mineral grains

▢ Clay matrix

▨ Natural cement

Sandstone *(below)*
Weathering accentuates the horizontal bedding planes and vertical joints in these sandstone masses.

Sandstones classified
Varying proportions of four components yield four types of sandstone, or arenite.

Components:
a Matrix
b Quartz
c Rock fragments
d Feldspar

Sandstones:
1 Greywacke
2 Quartzite
3 Lithic arenite
4 Arkose

Shale *(left)*
An outcrop reveals bedding planes in the fine-grained sedimentary rock, shale.

© DIAGRAM

4 | ROCKS FROM FRAGMENTS 2

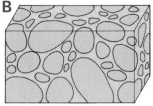

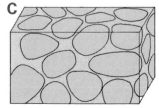

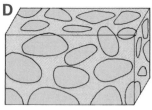

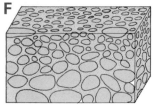

Rudites (from the Latin *rudis*, meaning "coarse") are clasts (rock fragments) coarser than a sand grain. Mixed with finer particles, rudites can be consolidated into natural concretes called conglomerates and breccias.

Conglomerates are named from the Latin for "lumped together." They contain rounded fragments —pebbles, cobbles, and/or boulders—and often represent waterborne and watersorted remnants of eroded mountain ranges or retreating rocky coasts. They accumulate along mountain fronts, in shallow coastal waters, and elsewhere, becoming mixed with sand, then bound by natural cement. How clasts in a conglomerate lie sorted, packed, and graded offers clues to how or where it was laid down. The thickest masses of conglomerate—as in the Siwalik Range, the southernmost part of the Himalayas' foothills—mark the aftermath of an orogeny.

Breccias (from the Italian for "rubble") are rocks containing sharp-edged, unworn, usually poorly sorted fragments, often embedded in a clay-rich matrix. Breccias usually form near their place of origin; their clasts have not been carried far enough to suffer rounding by abrasion. Many breccias originate in or from talus, deserts, mudslides, faulting, meteorite impact, or shrinkage of evaporite beds.

Authorities tend to separate conglomerates and breccias from tillites—poorly sorted, ice-eroded, ice-borne debris consolidated into solid rock. Many tillite clasts are faceted, with slightly rounded edges. Ancient tillites occur in South America, Africa, India, and Australia.

Types of conglomerate *(left)*
A Well sorted
B Poorly sorted
C Close-packed (clast supported)

D Loosely packed (matrix supported)
E Imbricated
F Graded bedding

Natural concretes

Main kinds of natural concrete.

1 Conglomerate (with rounded clasts)
2 Breccia (with sharp-edged clasts)
3 Examining an exposed breccia surface

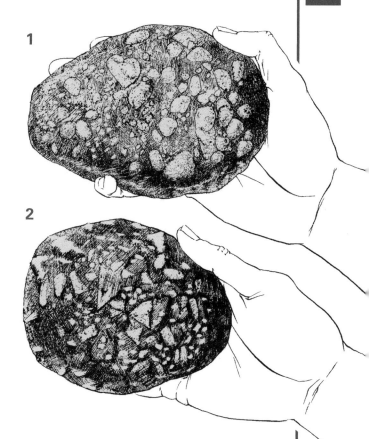

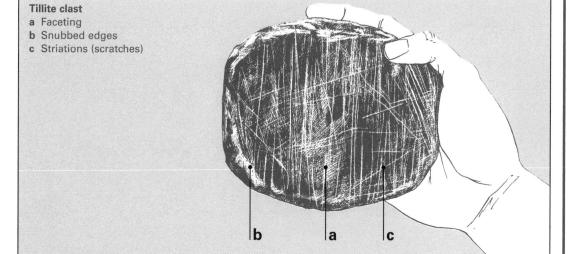

Tillite clast

a Faceting
b Snubbed edges
c Striations (scratches)

© DIAGRAM

4 | ROCKS FROM CHEMICALS

Oolites enlarged
This much-magnified section of a Cambrian limestone shows spherical oolites (also called ooids) cemented by carbonate. Each oolite's growth around a sand grain or shell fragment produced a concentric, radial structure.

Some sedimentary rocks and minerals consist of chemicals that had once been dissolved in water.

Certain limestones formed this way. Oolitic limestone consists of billions of oolites—tiny balls produced by calcium carbonate accumulating on particles rolled around by gentle currents in warm, shallow seas. Oolite forms like this today on the Bahama Banks. Dolomitic limestone (limestone mainly made of the mineral dolomite) occurs where certain brines chemically alter preexisting limestone or where dolomite deposits form in an evaporating sea.

Such so-called evaporites underlie one-quarter of the continents in beds up to 4,000 feet (1,220 m) thick. Evaporites now form where chemical deposits accumulate in evaporating desert lakes and coastal salt flats. But certain old evaporites could have been precipitated from chemically oversaturated deep offshore waters of almost landlocked seas such as the Mediterranean.

Three main minerals tend to settle in a sequence. First comes calcium carbonate. Next is gypsum (a granular crystalline form of calcium sulfate combined with water). Then comes sodium chloride in the form of halite (rock salt). This is a soft, low-density rock that is liable to flow. Pressure from overlying rocks forces up huge plugs or domes of salt beneath the coast of Texas and Louisiana and in parts of Germany, Iran, and Russia.

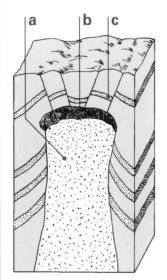

Salt dome (*above*)
a Salt core up to 6 miles (10 km) high
b Llimestone-anhydrite cap rock
c Strata uplifted by salt core

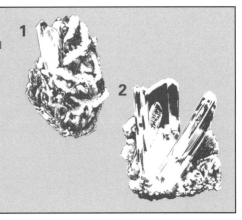

Two evaporites
1 Dolomite, derived from small shells, and named after the French geologist Dieudonné Dolomieu (1750–1801).
2 Gypsum, a sulfate mineral named from *gypsos*, the Greek word for chalk.

Besides the rocks and minerals just named, there are other chemical deposits. A few have or had important economic uses—particularly borax, chert and flint, certain iron-rich compounds, nitrates, and phosphorites. But some of these are partly biological in origin, and scientists disagree about how certain forms of iron occurred.

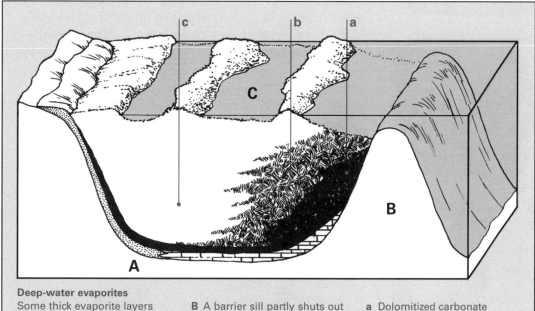

Deep-water evaporites
Some thick evaporite layers might have formed like this:
A In a deep water sea basin evaporation outstrips freshwater input.

B A barrier sill partly shuts out the open sea.
C The basin water grows denser and sinks until evaporites are precipitated in this sequence:

a Dolomitized carbonate
b Gypsum with anhydrite
c Halite (rock salt)

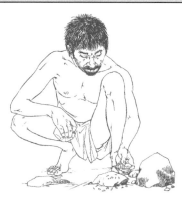

Flint knapping *(left)*
Stone Age humans learned that controlled blows could split this fine-grained quartz into sharp-edged weapons and tools.

4 | ROCKS FROM LIVING THINGS

Three organic rocks
1 Bituminous coal (mainly carbon)
2 Coquina (shelly limestone)
3 Chalk (a powdery limestone)

Organic sediments produce the rocks we know as coals and limestones.

Coals are rich in carbon derived from swampy vegetation. Coal type varies with the processes involved. Coal formation starts when plants die in wet acid conditions; instead of rotting completely, the plants turn into the soft, fibrous substance known as peat. Later, overlying sediments drive out moisture and squash the peat, converting it to lignite (soft brown coal). Even greater pressure gives bituminous coal—harder, blacker, and with a higher carbon content. The final stage is anthracite—a hard, black, shiny coal with the highest carbon content in the series. Most of the world's coal mines tap the remains of low-lying forests drowned by an invading sea and buried under sediments.

Coal forming
Coal forms from compacted dead plants. Increasing pressure boosts coal's carbon content and capacity to burn.
A Peat—the soft, moist, slow-burning product of dead plants in swampy land
B Lignite (brown coal) produced from squashed peat
C Bituminous coal—produced from squashed lignite; with only 3 percent water
D Anthracite—96 percent carbon and the richest source of heat from any coal

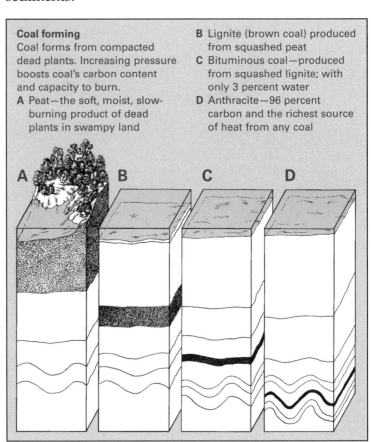

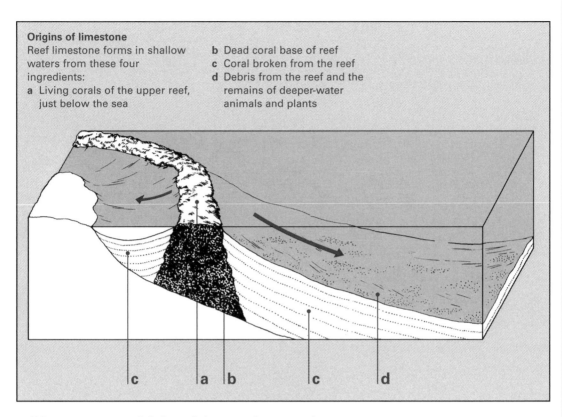

Origins of limestone
Reef limestone forms in shallow waters from these four ingredients:
a Living corals of the upper reef, just below the sea
b Dead coral base of reef
c Coral broken from the reef
d Debris from the reef and the remains of deeper-water animals and plants

c a b c d

Limestones are rich in calcium and magnesium carbonates. They make up about 8 percent of all sedimentary rock; only shale and sandstone are more plentiful. Organic limestones contain calcium carbonate extracted from seawater by plants and animals that used this compound for protective shells. These rocks include reef limestones built up from the stony skeletons of billions of coral polyps and algae inhabiting the beds of shallow seas. Coquina is a cemented mass of shelly debris. Chalk is a white, powdery, porous limestone made up of tiny shells of fossil microorganisms, drifting in the surface waters before they died and fell to the sea bottom.

CHAPTER 5 DEFORMED AND ALTERED ROCKS

Great loads of sediment or ice depress tracts of continental crust. Where a load has been removed, the land bobs up again. As lithospheric plates collide or split, tension or compression tilts, folds, squeezes, and breaks the rigid rocks, producing folds, faults, and earthquakes. Then, too, meteorites hurtling from space punch craters in the crust, while blobs of molten rock push up through the crust from below. Events like these produce heat and pressure that deform and alter solid rocks. This chapter ends with an account of rocks transformed that way.

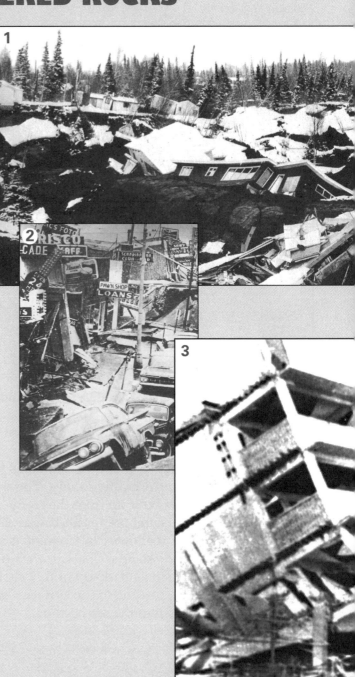

Scenes of destruction captured after earthquakes devastated Anchorage, 1964 (**1**, **2**), the southern Philippines, 1976 (**3**), and San Francisco, 1906 (**4–6**).

5 | RISING AND SINKING ROCKS

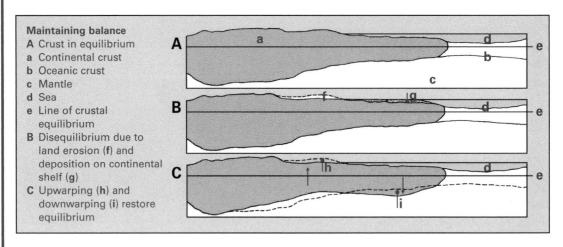

Maintaining balance
A Crust in equilibrium
a Continental crust
b Oceanic crust
c Mantle
d Sea
e Line of crustal equilibrium
B Disequilibrium due to land erosion (**f**) and deposition on continental shelf (**g**)
C Upwarping (**h**) and downwarping (**i**) restore equilibrium

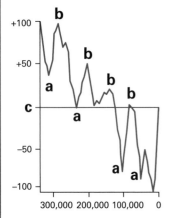

Changing sea level *(above)*
A graph depicts in meters the likely changes in the level of the Mediterranean in the last 300,000 years.
a Glacial periods
b Interglacials
c Present sea level

Great tracts of Earth's crust slowly rise or sink. Block mountains and the Black and Mediterranean seas are products of such movement, called epeirogenesis. Its cause is a type of disturbance that affects isostasy—the state of balance of Earth's crust floating on the denser underlying mantle. The crust is four-fifths as dense as mantle, so a mountain mass one mile high is "balanced" by four miles of crust below sea level.

Warping—the gentle rise or fall of crust—results from two pairs of processes: erosion and deposition, and freezing and melting.

Erosion that wears away a mountain mass reduces the weight pressing on the underlying crust, so the surface of the land bobs up. Such upwarping probably means that few large land surfaces get worn down to a level even with the sea. Indeed much of Africa is plateau, albeit perched on a superplume.

Deposition of eroded sediment depresses or downwarps deltas and tracts of the continental rim.

Sediments thousands of feet thick accumulate in eugeoclines—depressions in a continent's deepwater margin. Shallow-water sediments including limestones form long lenses in continental-shelf depressions called miogeoclines.

Growing ice sheets depress the crust beneath, squeezing out asthenospheric material, which pushes up the crust beyond the ice sheet's rim. When ice

A

a

d ↑   ↓ b   ↑ d

← c →

B

g ↓   ↑ e   ↓ g

→ f ←

Effects of ice
Simplified block diagrams show how an ice sheet makes land rise and sink.
A Ice sheet (**a**) depresses crust (**b**), displacing asthenosphere (viscous mantle) (**c**), which upwarps crust (**d**) around the ice sheet.
B After ice melts, depressed crust rises (**e**); displaced asthenosphere returns (**f**); and upwarped crust sinks (**g**).

sheets melt, the land below bobs up again—a process that has lasted 10,000 years or more in Scandinavia and northeast Canada.

A cross section through the Alps
This block diagram shows rock layers contorted by the collision of the lithospheric plates.

While isostatic change describes land moving up or down, eustatic change is a worldwide shift in the level of the sea. Eustatic change occurs if ocean basins grow or shrink, or if they gain or lose sea water. Sea level falls when much of the world's surface moisture becomes locked up in ice sheets on the land. But water freed by melting ice sheets lifts the level of the sea. Clues to eustatic and isostatic change include the drowned valleys and raised beaches of some coasts.

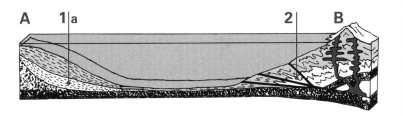

A   **1** a   **2**   **B**

Downwarped crust *(left)*
A section across a continental rim (**A**) shows a miogeocline (**1**) of shallow-water sediments colliding with a eugeocline (**2**) of deep-water sediments plus volcanic rocks generated above oceanic crust (**B**) where this is diving down beneath the continent into the mantle.

5 TILTING AND FOLDING ROCKS

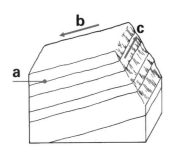

Tilted rocks *(above)*
The key features of tilted rock beds are:
a Bedding plane
b Dip
c Strike

Sediments and plateau basalts are laid down as horizontal beds or sheets. Old deposits lie almost undisturbed across great tracts of ancient, stable continental shields. But in unstable regions crustal tension and compression tilt, fold, squeeze, or break the level layers, and thrust them up as mountains. Which type of deformation happens depends on pressure, temperature, strain-rate (compression in a given time), and the composition of the affected rocks.

Folding occurs largely deep down along the edges of colliding continental plates. Here steady stress and high temperatures and pressures make normally brittle rocks bend instead of break. Thus in quartzite, quartz grains slide about and dissolve at stress points. Such processes produced repeated folding in the Alps and Himalayas.

Folded rocks exposed
This Cornish sea cliff reveals rock layers doubled over in a small recumbent fold. Rock folds of almost any kind can span just a few feet as here, or measure miles across, as in some regions of the Alps.

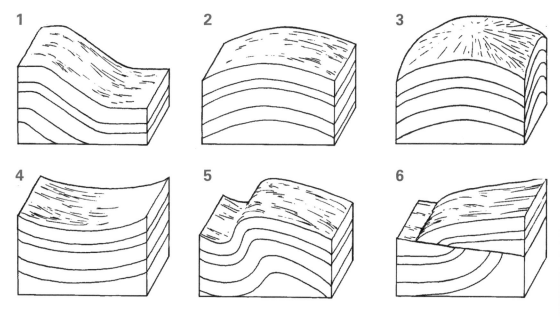

Folds take various forms, especially those above.

1 Monocline A steep steplike fold, bounded by upper and lower bends in a set of rock layers.

2 Anticline Rock beds upfolded into an arch from a few feet to many miles across. Anticlines form much of the Jura Mountains of France and Switzerland. An anticline containing many lesser folds is an anticlinorium. Huge anticlines are geanticlines.

3 Pericline An anticline in the form of an elongated dome. The Bighorn Mountains of Wyoming show a periclinal structure.

4 Syncline Downfolded sedimentary rock layers that form a basin such as the London Basin. A syncline with subsidiary synclines is a synclinorium. Immense synclines are called geoclines.

5 Overfold A lopsided anticline with one limb (side) forced over the other. Extreme overfolds are called recumbent folds.

6 Nappe A recumbent fold sheared through so that the upper limb is forced forward, perhaps for many miles. Nappes feature prominently in the Alps. Nappes with rocks forced over each other in slices like a pack of cards are imbricate structures.

Six Folds *(above)*
1 Monocline
2 Anticline
3 Pericline
4 Syncline
5 Overfold
6 Nappe

© DIAGRAM

5 | BREAKING ROCKS: JOINTS AND FAULTS

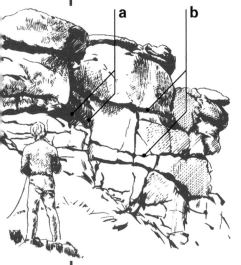

Joints and faults are splits that form in stressed rock, particularly near the surface.

Joints are cracks with little movement of the rock on either side. They open up as cooling igneous rock contracts, and appear in other rock, subjected to tension or compression. Joints occur in parallel sets, sometimes at right angles to each other.

Faults are breaks in Earth's crust, involving horizontal or vertical movement, or both, along a line of weakness called a fault plane. Block-faulting— breakup of a slab of crust into fault-bounded blocks— creates some of the world's great valleys and upland areas.

Joints (above)
In many sedimentary rocks, joints occur at right angles to bedding planes.
a Joints
b Bedding planes

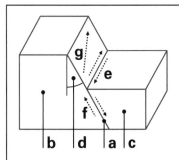

Anatomy of a fault (left)
a Fault
b Upthrow
c Downthrow
d Hade (inclination to vertical)
e Heave (lateral shift)
f Throw (vertical shift)
g Net movement

Rift Valley (below)
Such troughs lie between parallel faults with throws in opposite directions.

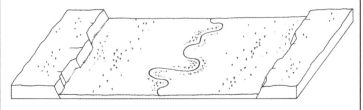

Here are six major types of fault or fault block.

1 Normal fault Stretching breaks rocks along a steep fault plane, and one block drops or rises against the other.

2 Reverse fault Compression forces one block up and over another. A thrust fault is a reverse fault with a low-angled fault plane producing great horizontal movement.

3 Tear fault (alias strike-slip, transcurrent, or wrench fault) Horizontal shearing along a vertical fault plane, as in California's San Andreas Fault. Transform faults are tear faults at right angles to oceanic ridges.

4 Graben A long, narrow block sunk between two parallel faults. Such blocks form the upper Rhine Valley, East African rift valleys, oceanic spreading ridges' central rifts, and other rift valleys.

5 Horst A horizontal block raised between two normal faults. Examples are the Black Forest, Vosges, Korea, and Sinai.

6 Tilt block An uplifted, tilted block.

Tilt-blocks form the western U.S. Basin and Range Province, Arabian and Brazilian plateaus, and the Deccan of India.

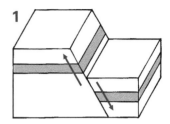

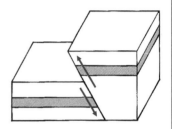

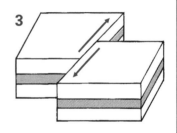

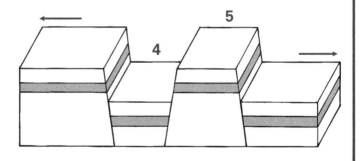

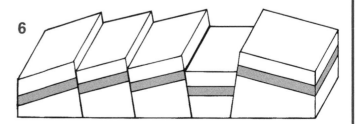

© DIAGRAM

5 EARTHQUAKES

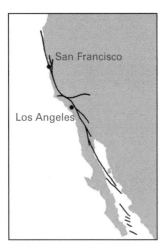

San Andreas Fault *(above)*
Lines show the San Andreas and associated faults. Land west of these is edging northwest.

An earthquake is a sudden shaking of the ground where stress-deformed rocks break along a fault and then snap back into shape, but in a new position.

An earthquake's point of fracture is its focus, which may be shallow, intermediate, or deep—down to about 430 miles (700 km). From the focus, two types of seismic wave pass through the rocks: Compressional waves (primary or P-waves) produce push-pull forces. Distortional waves (secondary, shear, or S-waves) are slower and make rock particles oscillate at right angles to wave direction.

Wave velocity increases with rock density and depth, and waves are reflected and bent on reaching the boundaries between two layers. But P- and S-waves differ in behavior. This helped scientists detect the Conrad discontinuity between upper "granitic" and lower, denser "gabbroic" crust; the Mohorovičić discontinuity between crust and upper mantle; and the Gutenberg discontinuity between the mantle and outer core (where S-waves cease because they cannot pass through the liquid outer core).

At the surface are long (L) waves subdivided into Love waves vibrating horizontally at right angles to their direction and Rayleigh waves that move through ground as waves move through the sea. By timing the arrival of these waves, three seismic stations can plot

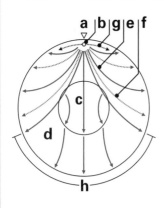

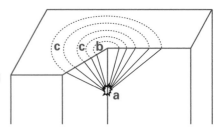

Seismic waves *(left)*
Waves reveal the Earth's core.
a Epicenter (above focus)
b Focus
c Core (blocks S-waves and deflects P-waves)
d Mantle
e P-waves
f S-waves
g L-waves
h Shadow zones

Earthquake shocks *(above)*
Isoseismal lines link places with equal intensity of shock.
a Focus
b Epicenter
c Isoseismal lines

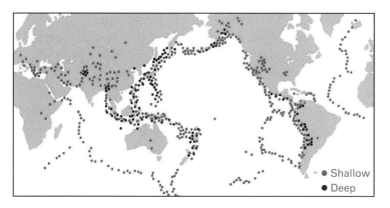

Earthquake belts *(right)*
These often coincide with active boundaries between lithospheric plates. Shallow earthquake foci lie 0–62 miles (100 km) down. Deep foci lie 62–435 miles (100–700 km) down.

the epicenter (the surface point above the focus).

L-waves spark off landslides, avalanches, and other earthquake damage. Undersea earthquakes set off tsunamis—mighty waves that devastate low coasts.

An earthquake's felt intensity is measured on the modified Mercalli scale, where 1 means "felt by few" and 12 means "damage total." The Richter scale measures magnitude, or energy released. Here each number stands for 10 times the energy of the number below; you would scarcely notice a 2, but an 8 would flatten a city. Most of the world's million earthquakes a year are fortunately slight.

Most occur at spreading ridges, oceanic trenches, and mountain-building zones. Strike-slip motion triggers earthquakes along California's notorious San Andreas Fault—a transform fault where western California moves northwest against the rest.

Evidence of earthquake
(below)
Past earthquakes may leave traces such as:
a River bend
b Displaced railroad
c Disrupted orchard

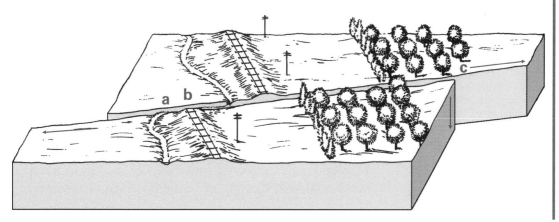

© DIAGRAM

BOMBS FROM SPACE

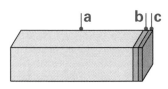

Meteorite percentages
(above)
A bar diagram shows relative abundance of the three main types of meteorite.
a Stony: 93%
b Iron: 5%
c Stony iron: 2%

Whizzing specks, stones, and rocks bombard the surface of Earth from space. About 10,000 short tons (9,000 metric tons) shower down each year. These missiles are meteorites—lumps weighing from a few ounces up to 100 tons or more. A few are bits of planets or the Moon struck off by other meteorites. Most are fragments of colliding asteroids—a belt of so-called minor planets between the orbits of the planets Mars and Jupiter. Some meteorites are 4,570 million years old, and their ingredients hold clues to the solar system's origin.

There are three main kinds of meteorite. Most are silicate-rich **stony meteorites**. The bulk of these are ordinary chondrites, containing solid silicate minerals. Carbonaceous chondrites include organic compounds, perhaps the building blocks that made life possible. Achondrites include basaltlike chunks, perhaps volcanic rock from big asteroids. **Iron meteorites**, the second major group, are iron with nickel. **Stony irons**, the third group, contain roughly equal amounts of nickel-iron and silicates.

Two meteorite types *(above)*
A Stony meteorite. Its white bits, perhaps from an exploding star, may be among the oldest solid matter in our solar system.
B Iron meteorite, etched and polished. Bands form a triangular pattern of nickel-iron and other nickel alloys.

Big stony meteorites break up and scatter before they hit the ground. Big iron meteorites vaporize upon punching impact craters in the ground. Circular depressions, shock structures in affected rocks, and tell-tale iridium and nickel deposits have helped scientists identify 90 sizable craters evidently gouged by meteorites. They include Canada's vast Manicouagan Crater, 43 miles (70 km) across; Arizona's famous Meteor Crater, 0.75 miles (1.2 km) across and

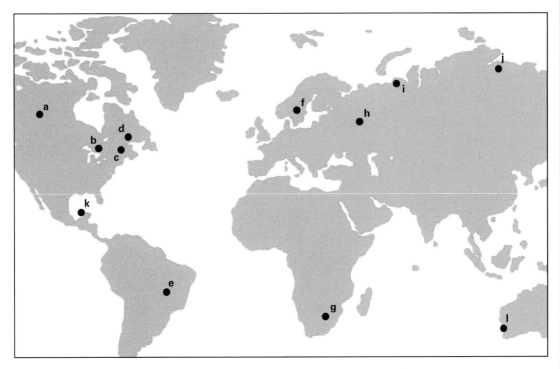

490 feet (150 m) deep; and Mexico's immense Chicxulub crater, 112 miles (180 km) from side to side.

More evidence for ancient impact lies in small, black, glassy buttons, spheres, and teardrops collectively called **tektites**. Found mostly in the Southern Hemisphere, tektites are blobs of molten sedimentary rock that splashed high above the atmosphere, before cooling, hardening and falling—widely scattered—back to Earth.

Impact craters
Dots mark 12 of the largest suspected craters—each more than 22 miles (36 km) across.
a Carswell, Canada
b Sudbury, Canada
c Charlevoix, Canada
d Manicouagan, Canada
e Araguainha Dome, Brazil
f Siljan, Sweden
g Vredefort, South Africa
h Puchezh-Katunki, Russia
i Kara, Russia
j Popigai, Siberia
k Chicxulub, Mexico
l Woodleigh, Australia

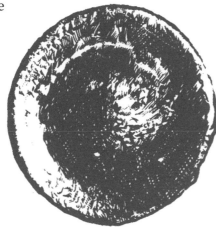

Tektite (right)
Such glassy "buttons" come from rocks melted and hurled high by meteorites punching craters like the one below.

ROCKS REMADE I

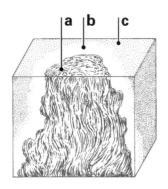

Contact aureole *(above)*
a Injected igneous mass
b Contact or metamorphic aureole of altered country rock
c Unaltered country rock

Great heat and pressure alter igneous and sedimentary rocks into metamorphic ("changed shape") rocks. These rocks' ingredients have undergone solid-state recrystallization to yield new textures or minerals. The greater the heat or pressure, the greater the change and the higher the grade of metamorphic rock produced. Here we summarize conditions that induce such changes. The next two pages describe the major types of metamorphic rock.

Metamorphism takes several forms. Burial metamorphism affects the base of immensely thick layers of sedimentary rock—the deeper the burial, the greater the pressure. Dynamic metamorphism transforms rocks crushed against each other in fault zones. Retrogressive (from high- to low-grade) metamorphism occurs at shear zones, for instance suboceanic transform faults where invading fluids introduce new elements that change the rocks' chemical composition—a process that is known as metasomatism. Impact metamorphism affects rocks struck by meteorites, but the best-known agents of change are contact and regional metamorphism.

Contact or thermal metamorphism occurs where a mass of magma invades and bakes country rocks (surrounding older rocks). Beyond a narrow baked zone extends a so-called contact aureole of altered rock.

Heat and pressure
This diagram relates rock-forming processes to temperature and pressure (indirectly shown by depth). Metamorphic rocks form in conditions between those producing sedimentary and igneous rocks.
a Diagenesis (sedimentary rock)
b Contact (thermal) metamorphism
c Burial metamorphism
d Regional metamorphism
e Anatexis or melting (igneous rock)

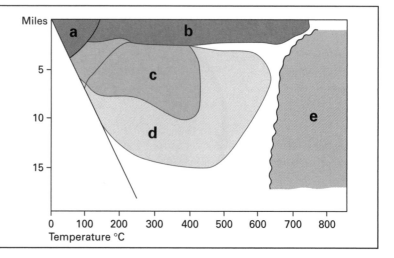

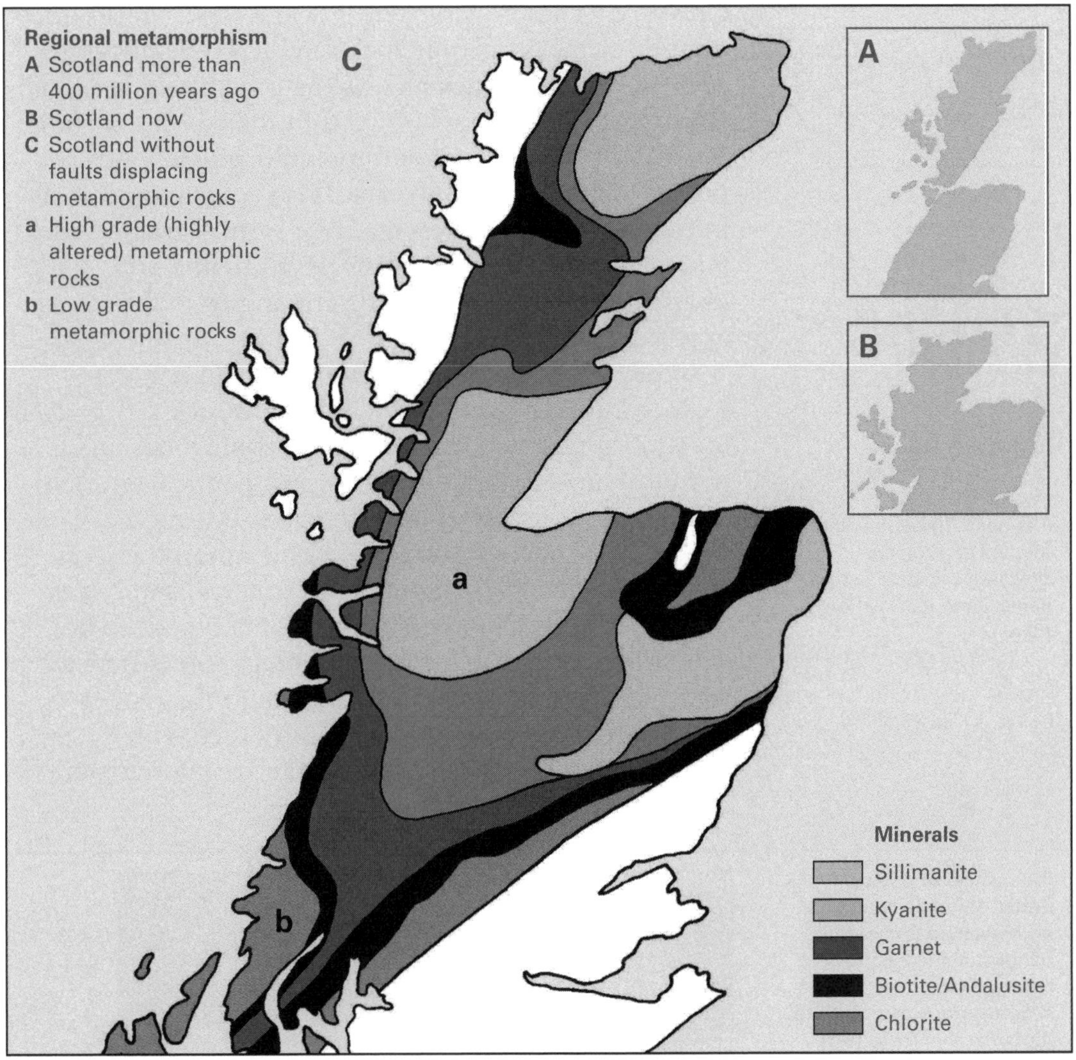

Regional metamorphism
A Scotland more than 400 million years ago
B Scotland now
C Scotland without faults displacing metamorphic rocks
a High grade (highly altered) metamorphic rocks
b Low grade metamorphic rocks

Minerals
- Sillimanite
- Kyanite
- Garnet
- Biotite/Andalusite
- Chlorite

An injected granite batholith may change the rocks for several miles around.

Regional metamorphism covers much larger areas, downwarped where colliding continental plates built mountains. Intensely altered rocks produced this way show up in the exposed roots of old, eroded mountain ranges, as in the Canadian Shield and parts of Scotland and Sweden, in the newer Alps and Himalayas, and in old subduction zones detectable in California, New Caledonia, and Papua New Guinea.

5 | ROCKS REMADE 2

1

2

Foliation *(above)*
1 Mica flakes haphazardly arranged in shale
2 The same flakes foliated— aligned by directed pressure—in slate, which splits along its foliation planes.

One type of sedimentary or igneous rock can produce a range of metamorphic rocks. Which type develops depends on such variables as the parent rock and the amounts of heat, pressure, and fluids passing through the rock. Contact (thermal) metamorphism tends to produce fine-grained textures. Heat plus pressure, as in regional metamorphism, favor coarse-grained rocks with foliated minerals—minerals flattened and aligned in parallel bands at right angles to the stress applied.

Index minerals—those formed at different temperatures or pressures—indicate a rock's metamorphic grade. Most metamorphic rocks are harder than sedimentary rocks, and pelitic (clay-rich) rocks undergo more change than basaltic rocks. We show eight common types of metamorphic rock: 1–4 formed largely by contact metamorphism; 2 and 5–8 by regional metamorphism. (Some types can be subdivided according to key minerals—for instance schist into talc schist and mica schist.)

1 Hornfels Fine-grained, dark, flinty rock with randomly arranged minerals; formed from mudstone and basalt.

From shale to gneiss *(left)*
Zones of intensifying heat and pressure (the parallel bands, read left-right) change surface belts of sedimentary rocks (**A–C**) to metamorphic rocks (**a–f**).
A Sandstone
a Quartzite
B Limestone
b Marble
C Shale
c Slate
d Phyllite
e Schist
f Gneiss

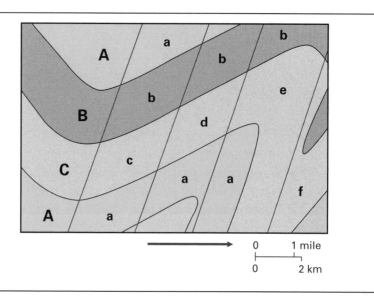

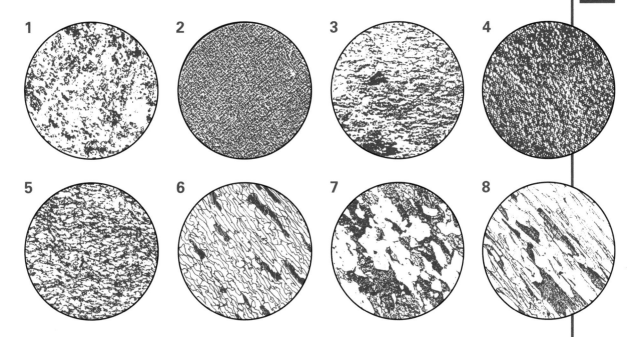

2 Slate Fine-grained, often gray, foliated rock split easily along cleavage planes of mica flakes aligned by pressure; formed from shale.

3 Marble Granular or sugary-textured rock; formed from limestone.

4 Quartzite Very hard, granular quartz rock; formed from sandstone.

5 Phyllite Silky, foliated rock more coarsely grained than slate, its usual precursor.

6 Schist Foliated rock, more coarsely grained and of higher metamorphic grade than phyllite; formed from slate or basalt.

7 Amphibolite Foliated rock of higher metamorphic grade than schist; formed from basalt.

8 Gneiss Foliated, banded, rock; coarser grained than schist and of the highest metamorphic grade.

CHAPTER 6 CRUMBLING ROCKS

As soon as rock is raised above sea level, weather starts to break it up. Water, ice, and chemicals split, dissolve, or rot the rocky surface until it crumbles. Crumbled rock mixed with water, air, and plant and animal remains form soil. Soil and broken rocks fall, flow, or creep downhill. These movements help create the slopes that form the surface of the land.

Chimney Rock on the Makinac waterway, Michigan, and cliffs in the Yellowstone National Park, Wyoming. (Engravings from *Picturesque America*, 1894)

6 ROCKS ATTACKED BY WEATHER 1

The wearing down of land begins as weather rots rocks at or near the surface. Different weather "weapons" probe different weaknesses; for instance, the joints in igneous rocks like basalt and granite, bedding planes in clays and shales, and natural cements in sandstones and conglomerates. Chemical weathering attacks chemical ingredients in rocks. Physical (mechanical) weathering destroys rock but leaves its chemicals unchanged. The kind of weathering that predominates depends upon the type of rock and climate.

Physical weathering is active chiefly in cold or dry climates. Agents include sharp temperature changes (which reinforce the chemical effects), frost, drought, crystallizing salts, and growing plants.

1 Exfoliation, or spheroidal weathering, is the flaking of intensely heated surface rock as it expands more than the cooler rock below. This process produces rounded, isolated rock masses called exfoliation domes.

2 Block disintegration involves sharp temperature

1 Exfoliation
Cutaway view of a boulder subject to exfoliation

changes, which make desert rocks expand and contract. This helps enlarge the joints in rocks, thus splitting large masses into smaller blocks.

3 Frost action Water expanding as it freezes widens crevices in well-bedded or well-jointed rock and shatters it; this occurs in winter in mid-latitude regions and at night in high mountains everywhere. Products are piles of sharp-edged debris, including cone-shaped slopes of talus (scree) seen below steep peaks. Freezing also causes the granular disintegration of porous rocks like chalk.

4 Tree roots can widen cracks in rocks as they grow.

5 Pressure release, or unloading, follows the removal of overlying rock and its pressure on the rock below. Expansion of that rock then forms curved joints promoting sheeting (pulling off) of rock shells from the inner mass.

6 Slaking is the crumbling of clay-rich sedimentary rocks as they dry out during drought.

7 Crystallization of salts Dissolved salts expanding as they dry and crystallize in rock split the rock and honeycomb its surface.

2 **Block disintegration**

3 **Frost action**

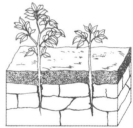

4 **Tree root action**

5 **Pressure release**
Yosemite Valley domes

© DIAGRAM

6 | ROCKS ATTACKED BY WEATHER 2

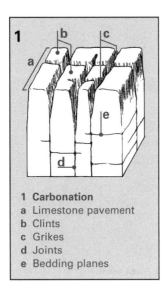

1 Carbonation
a Limestone pavement
b Clints
c Grikes
d Joints
e Bedding planes

Chemical weathering attacks rocks aggressively in humid climates. The chief destructive agents are rainwater and certain substances that it contains. These dissolve some kinds of rock or rot the natural cements that bind the particles in rocks together. But not all minerals are equally at risk; quartz proves much more resistant to attack than augite, biotite, hornblende, and orthoclase.

The following items summarize important ways in which chemicals make rocks decay.

1 Carbonation is the dissolving of limy rocks by percolating rainwater armed with carbon dioxide from the atmosphere or soil. The resulting weak carbonic acid widens joints in carboniferous limestone surfaces, producing bare limestone pavements where clints (sharp ridges) alternate with grikes (solution grooves). With items such as caves, swallowholes, and gorges these features form karst landscapes, named for a limestone region in Slovenia, former Yugoslavia.

Granite tor *(left)*
Such massively jointed tors crown hills in Dartmoor, England. The tors formed from groundwater weathering or possibly from Ice Age freeze-thaw weathering.

2 Hydration occurs when some minerals take up water and expand, thus breaking the shells from the rock containing them. Subsurface hydration probably produced southwest England's great, rounded granite moorland blocks called tors.

3 Hydrolysis is a water–rock reaction that can turn feldspar into clay, which then decomposes granite to produce white, powdery kaolin (china clay).

4 Solution occurs when water dissolves rock salt and (less readily) some other minerals.

5 Oxidation features the combination of atmospheric oxygen with the compounds in some rocks. Oxidized iron forms a brownish, crumbly, or (in dry lands) hard, protective crust of "rust."

6 Organic weathering is produced by the attack of organic acids from such organisms as bacteria, lichens, mosses, and decaying plants of many kinds on rock-forming minerals.

3 Hydrolysis
White mounds of waste quartz reveal a kaolin quarry, where rotting has changed and separated granite's mineral ingredients.

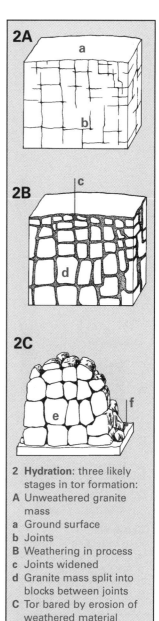

2 Hydration: three likely stages in tor formation:
A Unweathered granite mass
a Ground surface
b Joints
B Weathering in process
c Joints widened
d Granite mass split into blocks between joints
C Tor bared by erosion of weathered material
e Massive blocks exposed
f Lowered ground surface

6 SOIL FROM ROCK

In time most weathered rock acquires a covering of soil—a substance most land life depends on.

Soil forms as weathering breaks rock into particles ranging in size from clay to silt, sand, and gravel. Air and water fill the gaps between the larger particles. Then chemical changes help bacteria, fungi, and plants to move in. Plant roots bind groups of particles together; leaves ward off destructive rain; and roots and stems raise minerals from deep down. Plants and their remains form food for burrowing insects, worms, and larger creatures. Bacteria and fungi decompose dead plants, animal droppings, and dead animals, converting it all into dark, fertile humus. So soil's main ingredients are (inorganic) minerals, (organic) humus, air, water, and living organisms.

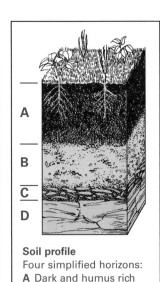

Soil profile
Four simplified horizons:
A Dark and humus rich
B Rich in minerals
C Infertile subsoil
D Unweathered bedrock

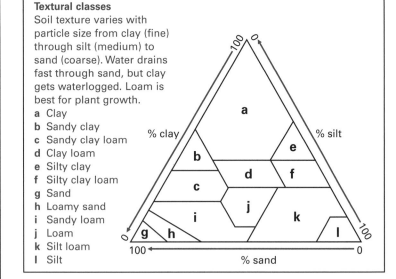

Textural classes
Soil texture varies with particle size from clay (fine) through silt (medium) to sand (coarse). Water drains fast through sand, but clay gets waterlogged. Loam is best for plant growth.
a Clay
b Sandy clay
c Sandy clay loam
d Clay loam
e Silty clay
f Silty clay loam
g Sand
h Loamy sand
i Sandy loam
j Loam
k Silt loam
l Silt

A slice cut down through soil reveals a profile made up of layers called horizons, from top to bottom known as A, B, and C. A is dark and rich in humus. B is rich in minerals and substances washed down from A, but paler, more compact and less fertile. C, the subsoil, consists of infertile weathered rock, derived from the unweathered bedrock (sometimes known as the D horizon).

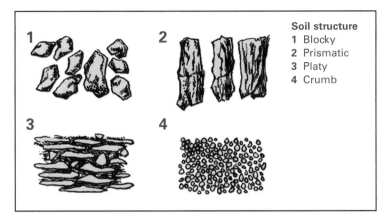

Soil structure
1 Blocky
2 Prismatic
3 Platy
4 Crumb

Few soils fit this description perfectly. Soil texture depends largely on the bedrock (parent rock), and soil type depends on topography, time, vegetation, and climate. Thus shales yield finer-textured soils than sandstones. Soils on limestones are rich in bases; others have an acid tendency. Soil depth can range from less than an inch (2.5 cm) on steep slopes to several yards on plains. Hillside soils are often better drained than those in valleys. Plants with different needs affect the proportions of some substances accumulating in the soil. But the chief influence on soils is climate.

Life in the soil *(left)*
These organisms live in or help to form the soil.
A Larger "aerators"
  a Mole
  b Earthworm
B Microorganisms
  c Fungus
  d Alga
  e Virus
  f Bacterium
  g Protozoan
C Arthropods
  h Wood louse
  i Millipede
  j Springtail
  k Cockchafer larva
  l Cricket
  m Ant
  n Mite

© DIAGRAM

6 TYPES OF SOIL

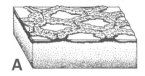

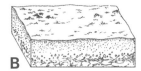

Six soil types *(above)*
Diagrams above show six soil types.
A Tundra soil
B Desert soil
C Chernozem
D Ferralsol
E Brown forest soil
F A red-yellow podzol

Pedologists (soil scientists) have many ways of classifying soils. Thus the Department of Agriculture labels soil types purely according to their properties, but the much-used zonal system stresses climatic origins.

Soils certainly owe more to climate and vegetation than to bedrock. Heavy rainfall causes much leaching (downward flow of dissolved substances), eluviation (downward movement of fine particles), and illuviation (redeposition of these substances at lower levels), and so it affects soil fertility. At the other extreme, hot deserts undergo salinization or alkalization as soil water brings salts and alkalis to the surface, then evaporates, leaving the chemicals to form a whitish crust.

Most of the following soil types come from the zonal system. The Arctic's tundra soils are often waterlogged or frozen, with a peaty upper layer and bluish mud below. The cool-climate podzol ("ash") of northern coniferous forests is acid with a leached, ashy B horizon above a hard thin illuvial pan. Temperate forests of the world produced brown forest soil—humus-rich and slightly acidic. Temperate grasslands (the steppe, prairie, pampas, and Australian downs) include chernozem, and/or prairie soil, and chestnut-brown soil. Chernozem's dark, humus-rich upper layer formed under light rainfall with little leaching. Prairie soil and chestnut-brown soil formed in drier climates. Alfisols ("degraded chernozems") have been identified in Spain, northwest Africa, India, and parts of Africa. Hot deserts have pale, coarse, soils, poor in humus and sometimes white with salty crust. Tropical grasslands include dark, clayey grumusols. Deep, reddish, iron rich ferralsols underlie the humid tropics. Mountain soils include thin scree soils—little more than rock fragments.

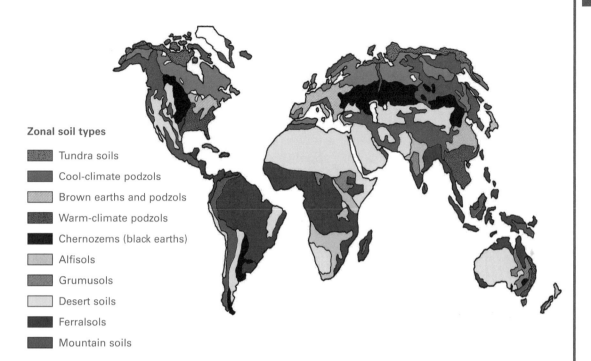

Zonal soil types

Tundra soils

Cool-climate podzols

Brown earths and podzols

Warm-climate podzols

Chernozems (black earths)

Alfisols

Grumusols

Desert soils

Ferralsols

Mountain soils

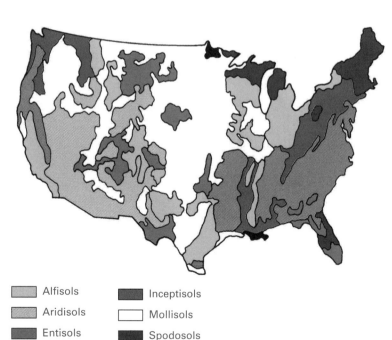

Soils of the United States
This map uses the Department of Agriculture's classification.
Alfisols: leached basic soils
Aridisols: salty desert soils
Entisols: young mineral soils
Histosols: organic soils
Inceptisols: immature soils
Mollisols: soft soils with a richly organic top layer
Spodosols: leached acidic soils
Ultisols: leached, much weathered acid soils

Alfisols

Aridisols

Entisols

Histosols

Inceptisols

Mollisols

Spodosols

Ultisols

© DIAGRAM

6 MASS MOVEMENT

Types of mass movement
Numbered block diagrams refer to numbered text items.

1 Soil creep with (**a**) bent weathered rock and (**b**) displaced trees, rocks, etc.
2 Earth flow with (**c**) stepped slope and (**d**) downslope "toe"
3 Mudflows issuing from canyons in a desert
4 Slumping with (**e**) sandstone slump block after sliding down (**f**) a cliff of weak shale
5 Rockslide showing (**g**) scar torn in mountainside and (**h**) landslide debris
6 Rockfalls producing (**i**) scree or talus cones of debris fallen from a cliff (**j**)

Mass movement (mass wasting) is the force of gravity shifting weathered rock and soil (the regolith) downhill. Water often lubricates and aids this process. Amounts and speeds involved vary with such things as slope steepness, underlying rocks, and quantity of moisture in the ground. Mass movement can affect a few square yards or a whole mountainside. Sodden regolith may flow; dry regolith will slide or fall. Slow movements include creeping and flowing. Swift movements include landslides and rockfalls triggered by such things as earthquakes, heavy rain, or quarrying.

1 Soil creep is the imperceptible downslope creep of soil, betrayed by items such as tilted and displaced trees and fences.

2 Earth flow This features a stepped slope where material has slumped downhill and a bulging downslope "toe" where it accumulates. Earth flow can occur in hours on saturated slopes.

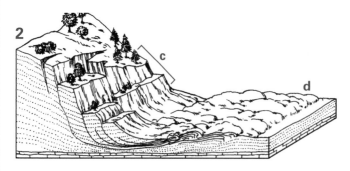

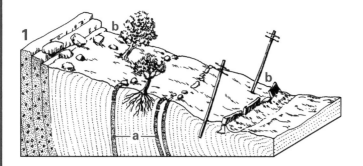

3 Mudflows occur where mud holds so much water that it flows as slurry even down a gentle slope. A mudflow containing volcanic ash and dust from Mt. Vesuvius swallowed the Roman city of Herculaneum in AD 79.

4 Slumping is a landslide where rock masses tilt back as they slide from a cliff or escarpment—often where well-jointed sedimentary rocks overlie clay or shale. Resulting slump blocks can be 2 miles (3 km) long and 500 feet (150 m) thick.

5 Rockslides involve masses of bedrock slipping down a sloping fault or bedding plane. Rockslides have killed people and destroyed villages in the Canadian Rockies, Norway, and Switzerland.

6 Rockfalls are free falls of fragments of any size from a cliff. Frost-shattered fragments in time form cliff foot talus cones, with a talus (or scree) slope at an angle of about 35 degrees.

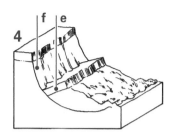
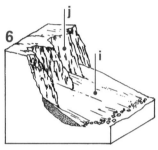
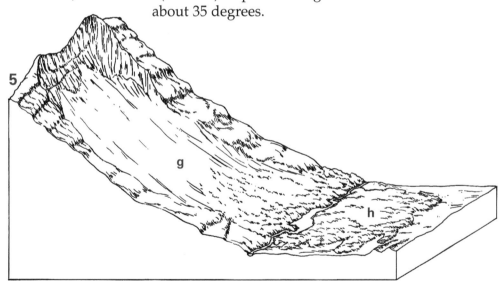

SLOPES

Slope segment angles
Block diagrams show slope steepness influenced by input and output of (**a**) weathered debris overlying (**b**) bedrock.
1 Graded slope: Debris output equals input with none added by the segment.
2 Graded slope: Output equals input including debris added by the segment.
3 Steep slope: Input exceeds output.
4 Gentle slope: Output exceeds input.

Slopes steep or gentle make up almost all the surface of the land. Everywhere, mass movement, splashing raindrops, or rainwater flowing over land are forming slopes and wearing them away by shifting soil or broken bits of rock downhill.

Wherever the force of gravity is greater than the force of friction holding particles upon a slope, these tend to slide downhill. Most slopes have an average angle of less than 45 degrees. But a single slope usually has several (straight, concave, or convex) segments— parts with different angles. Slope angle varies with the amount of weathered debris entering and leaving a segment.

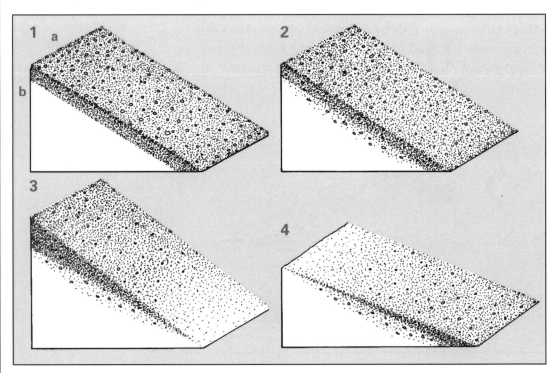

Slopes probably evolve in three main ways, all possible in one location:

A Slope decline Slope angle decreases through stages: (**a**) steep free face; (**b**) graded slope with convex curve above, concave curve below; (**c**) decreasing curvature; and (**d**) reduction in height. Slope decline predominates in moist temperate areas such as the northeastern Appalachians.

B Slope retreat The retreating slope keeps a short convex top, long free face, debris slope, and (lengthening) pediment (thin sheet of debris). Such slopes abound in semiarid areas.

C Slope replacement Lower angle slopes extend upward to replace steep upper segments.

A

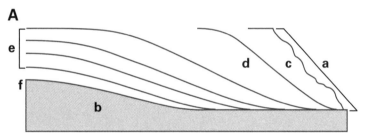

A Slope decline
a Former position of slope
b Slope's present position
c Steep free face
d Graded slope
e Decreasing curvature
f Reduced height

B

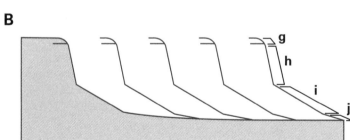

B Slope retreat
g Short convex top
h Long free face
i Debris slope
j Pediment

C

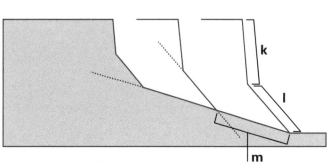

C Slope replacement
k Steepest slope, replaced by:
l Gentler slope, replaced by:
m Gentlest slope

© DIAGRAM

CHAPTER 7 HOW RIVERS SHAPE THE LAND

Rain-fed rivers carve valleys in the hills and wash soil and weathered rock downhill, dumping them into the lakes and seas. Year by year the erosive work of rivers wears down the highest mountain ranges until some form flat plains that barely peek above the sea. Off some sheltered shores, though, sediments shed by rivers build deltas—muddy aprons of new land. This chapter describes the work of rivers and the landforms many help create.

The Yellowstone River Valley, Wyoming. (Engraving from *Picturesque America*, 1894)

RUNNING WATER

Soon after land appears above the sea, rivers set about attacking it. Rivers rise in highlands and flow downhill to empty in a sea or inland drainage basin. The force of their moving water erodes and transports a load of soil and rock, so carving valleys that dissect mountains into peaks and ridges and reducing these to hills. But rivers also deposit the eroded debris to build lowland plains and offshore underwater platforms.

In theory, by degrading (eroding) some stretches of its bed and aggrading (building) others, a river tends to gain a graded concave profile, leaving it with just enough velocity to shift its load. In theory, too, rivers tend to bevel continents to peneplains—almost level plains just above sea level. In practice, earth movements and differences in rock formation and resistance to erosion interrupt both trends.

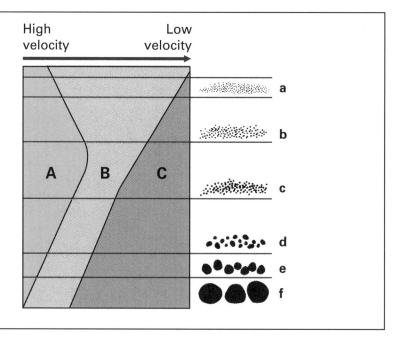

Water transportation
Stream velocity affects differently sized particles in different ways. Here we show how velocity decrease (from left to right) corresponds to erosion (**A**), transportation (**B**), and deposition (**C**) of the following:
a Clay
b Silt
c Sand
d Pebbles
e Cobbles
f Boulders

But calculations confirm running water's power to sculpt the continents. Rivers drain about 70 percent of all dry land. They contain only 0.03 percent of all fresh water, yet carry enough each year to drown all dry land 1 foot (30 cm) deep. Every year rivers dump about

20 billion tons of eroded material in the sea—enough to shave 1.2 inches (3.13 cm) off Earth's land surface every thousand years. The Mississippi River alone shifts an estimated 516 million tons a year; 340 million tons as fine suspended particles, 136 million tons in solution, and 40 million tons by saltation—the hopping of heavy particles along the river bed.

An individual river's power to erode and transport depends largely on its discharge—the product of its volume and velocity. The greatest mean discharge is the Amazon's 6,350,000 cubic feet per second (180,000 m³ per second)—nearly ten times the Mississippi's. But for rivers such as China's Yangtze, wet and dry seasons affect the flow from month to month.

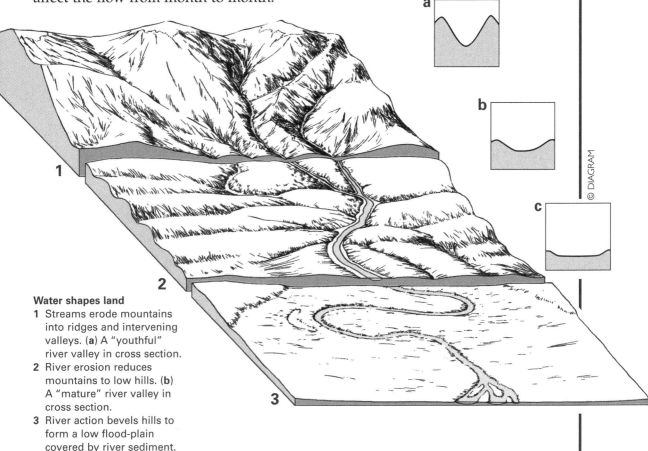

© DIAGRAM

Water shapes land
1 Streams erode mountains into ridges and intervening valleys. (**a**) A "youthful" river valley in cross section.
2 River erosion reduces mountains to low hills. (**b**) A "mature" river valley in cross section.
3 River action bevels hills to form a low flood-plain covered by river sediment. (**c**) An "old" river valley in cross section.

7 | WATER COMES AND GOES

The chief agent that wears down dry land is water flowing on the surface as part of the water cycle. Powered by the Sun and gravity, the cycle starts as the Sun's heat evaporates surface water, mostly from the oceans. Water vapor in the atmosphere cools and condenses into droplets that build clouds. Clouds shed this moisture as rain or snow. Much of this precipitation falls on the sea, but some falls on dry land. Vast quantities are locked in slowly moving sheets of ice. But some water seeps underground, and smaller quantities run off the surface as rills, converging in rivers and lakes. Most water returns to the sea. Rainfall is occasional, but many rivers are perennial, fed by underground water, which includes the following sources:

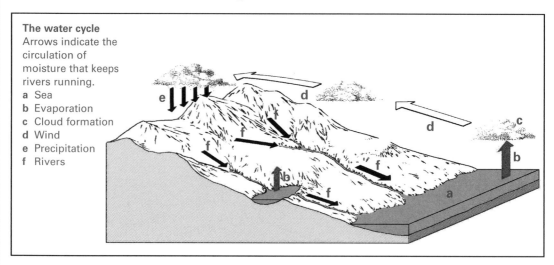

The water cycle
Arrows indicate the circulation of moisture that keeps rivers running.
a Sea
b Evaporation
c Cloud formation
d Wind
e Precipitation
f Rivers

1 Ground water Below the soil, ground water saturates permeable rocks (rocks that let water through), filling the pores of porous rocks, such as sandstone, and cracks in pervious rocks, including limestone.
2 Aquifer This is a saturated layer of permeable rock lying on a layer of impermeable rock, such as slate or shale. The aquifer's surface is its water table. The table's level varies with rainfall, tends to follow slopes up and down, and is exposed in swamps, lakes, and springs.

3 Spring This is a flow of water escaping from the ground, such as where the water table outcrops on a hillside above impermeable rock. Springs are the sources of some major rivers.

4 Artesian basin This is a saucer-shaped aquifer sandwiched between layers of impermeable rock. Rain soaks down through its rim. Below rim level, water under pressure may gush from artesian wells and artesian springs. Artesian basins underlie London, Paris, much of the Sahara Desert, Australia, and North America.

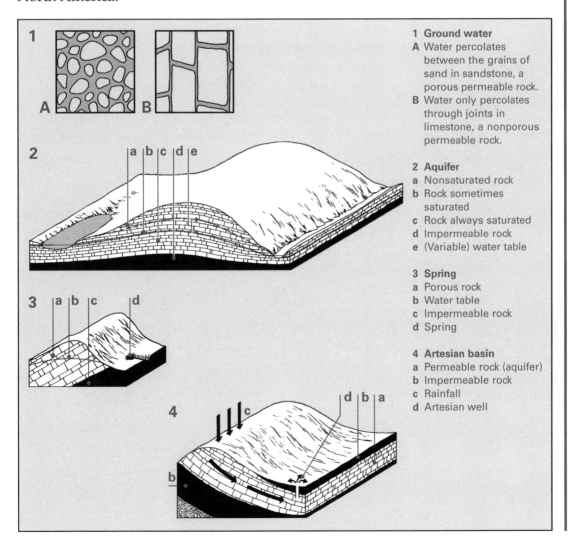

1 Ground water
A Water percolates between the grains of sand in sandstone, a porous permeable rock.
B Water only percolates through joints in limestone, a nonporous permeable rock.

2 Aquifer
a Nonsaturated rock
b Rock sometimes saturated
c Rock always saturated
d Impermeable rock
e (Variable) water table

3 Spring
a Porous rock
b Water table
c Impermeable rock
d Spring

4 Artesian basin
a Permeable rock (aquifer)
b Impermeable rock
c Rainfall
d Artesian well

© DIAGRAM

7 HOW RIVER VALLEYS FORM 1

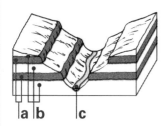

Resistant strata (above)
Alternating resistant and easily eroded horizontal strata help to produce river valleys with a stepped cross profile.
a Resistant strata
b Easily eroded strata
c River

A mountain valley (below)
a Gully
b Tributary stream
c Alluvial fan deposited by tributary
d River
e Divide

River valleys grow where rivers cut down and sideways into rock, and weathering wears back the slopes on either side. Steep-sided valleys and broad, flat-bottomed valleys were once seen as erosion-cycle stages produced, respectively, by youthful and mature sections of a river. This takes no account of the effects of local rocks and climate. But upper, middle, and lower reaches of a river do often show distinctive differences.

Many valleys start to form high up on slopes, where springs erupt or rainwater successively produces splash, sheet, then rill erosion; deepened rills cut river channels. In its upper course a river may be a small, fast-flowing torrent cutting down into its bed and forming rapids and waterfalls. Floods accelerate this process, transporting rocks that gouge out rock pools. Mountain streams commonly reveal these features:

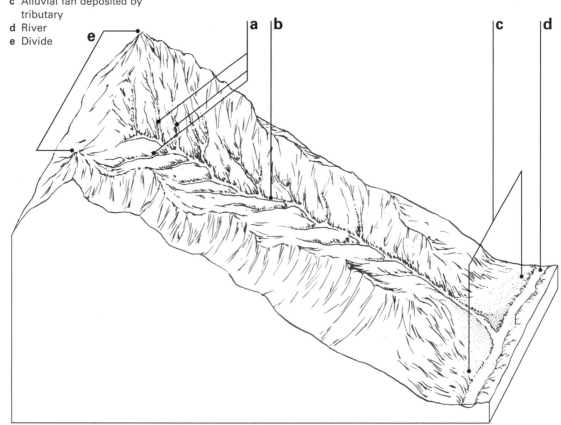

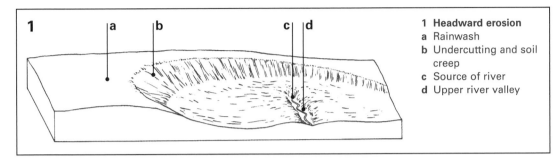

1 **Headward erosion**
a Rainwash
b Undercutting and soil creep
c Source of river
d Upper river valley

1 Headward erosion Undercutting, rainwash, and soil creep help a river valley to gnaw back into a hillside, thus lengthening the river.

2 Pot-holes Circular holes in a rocky stream bed show where waterborne scraps of rock were whirled around, deepening depressions. Stream-channel erosion also rubs bits off the rock fragments themselves, a process called attrition.

3 V-shaped valley Valleys develop deep, steep cross sections where streams erode downwards. Interlocking spurs are ridges projecting from both sides of the valley. The stream erodes most on the outsides of the bends, where the current flow is strongest, so the zigzags grow more pronounced.

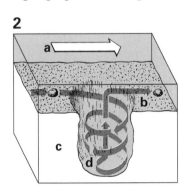

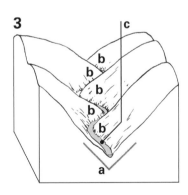

2 **Pot-hole**
a Stream flow
b Path of pebble
c Stream bed
d Pot-hole

3 **V-shaped valley**
a V-shaped cross section
b Interlocking spurs
c Stream

7 | HOW RIVER VALLEYS FORM 2

A river valley's middle section typically shows more mature features than its upper course. Weathering has broadened the valley sides. The river flows down a gentler gradient. Its current travels fast enough to shift the load of sediment acquired upstream. But the stream flows mostly over sediment it has deposited and no longer cuts down into the rock below its bed. Rather, it erodes from side to side, nibbling away its banks, thereby flattening and broadening the valley floor. Yet even in this middle section, land uplift or an outcrop of resistant rock may interrupt the valley's gently curving profile with a waterfall. The illustrations show these processes at work.

1 Meander This is where a river current flows around a bend, hits its concave bank at speed, and gnaws that bank away to form a river cliff. Meanwhile, the river deposits sediment in slack water on the convex bend, where a sloping spur known as a slip-off slope grows out into the river. Thus the river migrates slowly sideways.

1 Meander
This diagram of a meander (marked curve in a river channel) shows these features.
a Erosive surface flow
b Concave bank
c River cliff
d Flow on river bed, shedding sediment
e Convex bank
f Shingle
g Slip-off slope

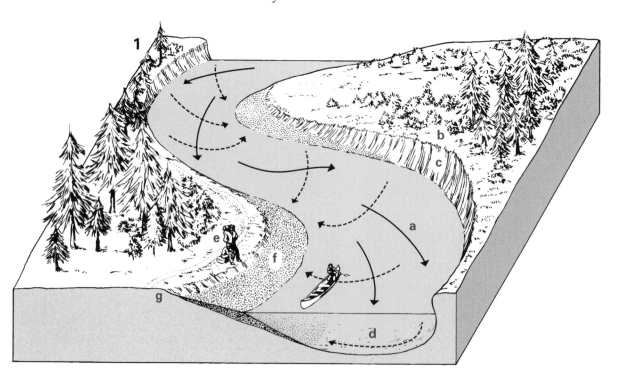

2

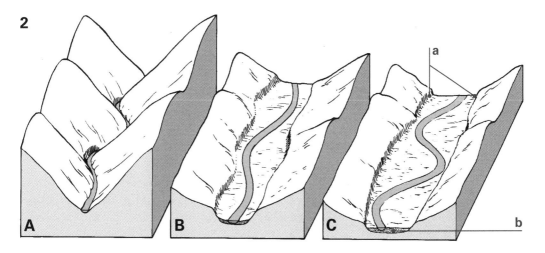

A   B   C   a   b

2 Valley broadening Meanders slowly migrate downstream, widening the river valley by lopping off the ends of interlocking spurs. This rims the valley floor with low cliffs called bluffs.

3 Waterfall This is a clifflike face down which a river plunges. Waterfalls form where resistant rock or land uplift interrupt the river's profile. Major falls include the Congo River's Boyoma (Stanley) Falls, the Paraná's Guaíra Falls, Niagara Falls, and the Zambezi's Victoria Falls. Vertical river erosion cuts back some waterfalls, so they migrate upstream above a long, deep gorge.

2 Valley broadening
A sequence of three diagrams shows migrating meanders widening a valley floor.
A Lateral erosion starts at concave banks.
B Lateral erosion lops off spurs.
C Large meanders migrating downstream broaden the valley floor, carving bluffs (**a**) and dumping gravel (**b**).

3A **3B**

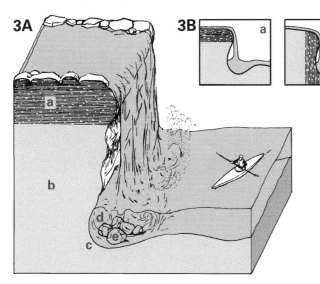

a   b   c

3 Waterfalls
A Waterfall features:
a Resistant rock
b Weaker rocks
c Undercutting
d Plunge pool
e Boulders
B Resistant rocks producing waterfalls:
a Horizontal cap rock
b Vertical rock
c Gently sloping rocks (creating rapids)

© DIAGRAM

7 | WHERE RIVERS SHED THEIR LOADS

In their lower courses, huge rivers like the Amazon and Mississippi flow down a gradient as slight as 3 inches per mile (5 cm per km). Here, the rivers have beveled off all hills and cross a broad, low-lying plain thickly carpeted with sand and mud. Vast quantities of these river sediments built the floodplains of the Ganges, Mississippi, Niger, Nile, Rhine, and Rhône. Deposition blocks such rivers' mouths, producing deltas—swampy plains through which the river flows, divided into several channels called distributaries. Deltas grow where a river sheds a large load of sediment faster than tides and currents can carry it away.

The illustrations show floodplain and delta features.
1 Floodplain This has a valley floor flattened and broadened by river erosion and floored with sediments deposited by migrating meanders and river floods.

1 Floodplain
These features appear in a river's floodplain:
a Snaky meanders
b Levees
c Oxbow lakes
d Mud, silt, and sand
e Bluffs

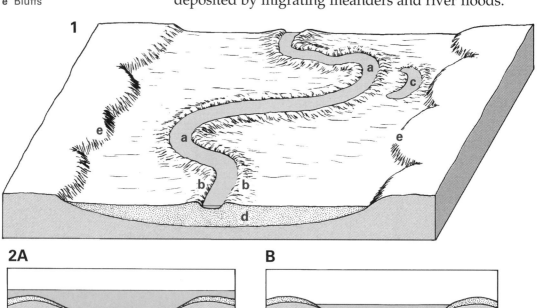

2A

B

C

2 Levees
Cross sections show how a river raises its banks.
A Sediment deposited by flooding
B Sediment deposited in normal flow
C Sediments and river after repeated flooding

Cutoff meanders survive as oxbow lakes. The river flows above the level of the plain and between raised banks called levees.

2 Levees form because the river floods repeatedly. A flooding river sheds mud on its banks where current flow is slow, so banks grow higher. After flooding, the river deposits sediment on its bed, which raises it. After repeated flooding, both bed and banks are raised, and the river surface lies higher than the plain on either side.

3 Delta Deltas take the shape of a fan or bird's foot. They form in stages. Deposition splits a river into distributaries flanked by levees and separated by lagoons. Sediments turn lagoons into swamps and build bars and spits. Swamps become dry land (Louisiana, Mesopotamia, and the North China Plain were formed from deltas in this way).

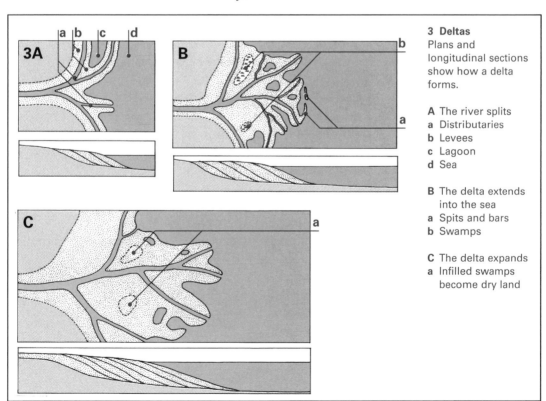

3 Deltas
Plans and longitudinal sections show how a delta forms.

A The river splits
a Distributaries
b Levees
c Lagoon
d Sea

B The delta extends into the sea
a Spits and bars
b Swamps

C The delta expands
a Infilled swamps become dry land

© DIAGRAM

RIVERS REVIVED

7

1A

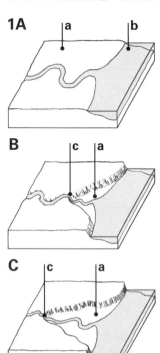

1 Knickpoint
Block diagrams and long profiles show a knickpoint receding upstream from a river mouth.
A Before land uplift
B Soon after uplift
C Later
  a Floodplain
  b Sea
  c Knickpoint

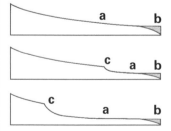

Mature, meandering rivers sometimes start vigorously cutting down into their beds like youthful mountain torrents. Called rejuvenation, this process happens if a lasting rise in rainfall boosts a river's flow, or if land uplift or a fall in sea level leaves the river mouth above the sea. Either way the river will regrade its bed. Various landforms show where this has happened.

1 Knickpoint This is a sharp step in a river's long profile, often marked by rapids or a waterfall. In time a knickpoint starting at a river's mouth recedes upstream. Some rivers show several knickpoints resulting from successive uplifts of the land.

2 River terraces are remnants of an old floodplain left when a river cuts down and sideways in the sediments through which it flowed before rejuvenation. Renewed rejuvenation cuts a second pair of terraces below the level of the first. Further episodes of uplift produce more terraces. Several pairs flank the Thames in London.

3 Entrenched (incised) meanders are floodplain meanders incised in bedrock by rejuvenation of a river. The Goose Necks of the San Juan River in Utah have been cut down through horizontal beds.

4 Gorges and canyons are deep, steep-sided, rocky valleys cut by rejuvenation in resistant rock or along a fault. Land uplift helped produce the 3-mile (4.8 km) deep Himalayan gorges of the Ganges and Brahmaputra, and Arizona's 1.5-mile (2.4 km) deep Grand Canyon—a desert gorge cut by the Colorado River, which is fed by distant melting snows.

5 Natural bridge This type of rock formation is formed by cutoff of an incised meander. Rainbow Bridge at Navajo Mountain, Utah, is a striking rock bridge rising 309 feet (94 m) from a gorge floor and spanning 278 feet (85 m).

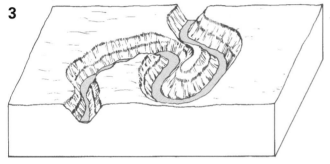

2 River terraces
a Oldest terraces
b Younger terraces
c Present floodplain
d Present knickpoint
e Sediments

3 Entrenched meanders
River erosion has kept pace
with uplift of a floodplain by
deeply incising old meanders
in the rising land.

4 Gorges
River rejuvenation cut this
long, steep, narrow gorge in
land rising as earth
movements built mountains.
a Mountain
b Gorge

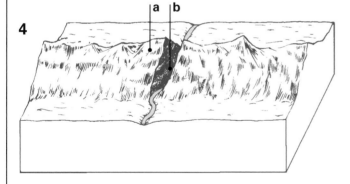

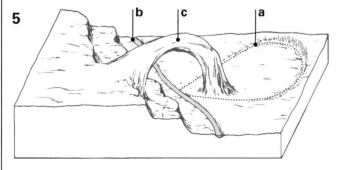

5 Natural bridge
Lateral river erosion whittled
away cliffs flanked by an
incised meander. Then the
stream cut through the
meander's narrow neck.
a Old incised meander
b Present course of river
c Natural bridge

7 | RIVERS UNDERGROUND

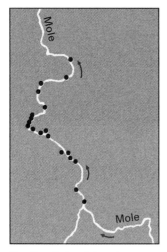

Swallowholes *(above)*
Dots show swallowholes in the chalk bed of England's River Mole. Water tends to vanish down these holes in dry weather.

How caves form *(right)*
Three illustrations trace the growth of caves in limestone.

A Rainwater trickles down through crevices produced by:
a Joints
b Bedding planes

B Rainwater acting as a weak carbonic acid dissolves the rocks it touches and removes the dissolved material. This widens vertical and horizontal crevices in limestone.

C Streams plunging underground widen crevices into vertical and horizontal caves. If the climate changes so that rainfall drops, many caves are left quite dry.

In most landscapes, rivers flow across the surface of the land. But rainwater sinks down through joints in permeable limestone rocks and invisibly attacks them underground. Carbon dioxide gas combines with falling rain to turn the water into weak carbonic acid, and this carbonic acid dissolves calcium carbonate—the main ingredient of limestone. In moist chalk countryside this process removes an estimated 35 tons of rock a year from every acre and forms depressions called solution hollows. In carboniferous limestone, water enlarges vertical and horizontal joints to gnaw complex channels underground. A slice cut through some limestone mountains would resemble a giant slice of Swiss cheese. Some limestone caverns and cave systems are immense. Several European cave systems descend 4,000 feet (1,219 m) or more. Mapped passages in Kentucky's Mammoth Cave National Park exceed 230 miles (370 km). Borneo's Sarawak Chamber could hold more than 7,000 buses, and five football fields would fit in the Big Room of New Mexico's Carlsbad Caverns.

A |a |b

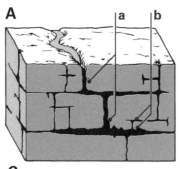

B

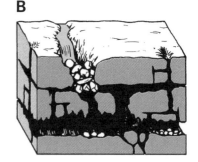

C

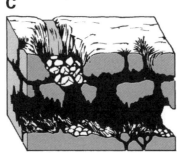

Where dripping water evaporates and/or gives up carbon dioxide, the dissolved calcium bicarbonate becomes insoluble in the mineral form called calcite, building deposits such as these:

1 Stalactites "Icicles" of calcite hanging from a cave ceiling. Depending on conditions they take 4 to 4,000 years to grow 1 inch (2.54 cm).

2 Stalagmites Calcite spikes jutting upward from a cave floor.

3 Columns Calcite forms produced where stalactites and stalagmites meet.

4 Gours Calcite ridges formed where water rich in carbonate flows over an irregular surface.

A limestone cave system
a Clints (blocks)
b Grikes (gullies)
c Sink-holes
d Galleries
e Stalagmites
f Stalactites
g Columns
h Gours

© DIAGRAM

7 | DRAINAGE PATTERNS

Parallel pattern
Parallel tributaries join at an acute angle. This pattern is typical of young river systems on uniform rock.

1 Dendritic pattern
Small branching tributaries flowing on uniform rock unite to form a network like the twigs, branches, and main trunk of a tree.

2 Trellis patten
a Consequent stream (following the land's original slope)
b Subsequent stream (at right angles to consequent stream)
c Obsequent stream (opposite in direction to a consequent)

Land drained by a river and its tributaries comprises a watershed, also called a drainage basin, river basin, or catchment area. Some watersheds are immense. The Amazon's, the world's largest, embraces almost two-fifths of South America.

A drainage basin's river channels form a drainage pattern that depends on slope, rock types and formations, and crustal movements. Drainage patterns include the following types and features.

1 Dendritic pattern Named from the Greek *dendron*, or "tree," this is a tree-shaped pattern of small, branching tributaries feeding into a main "trunk" river. It forms on rocks of equal resistance.

2 Trellis pattern Here some tributaries flow parallel to the main river, others at right angles to it. This pattern appears where bands of hard and soft rock alternate.

3 River capture This involves a stream eroding headward until it captures another stream's headwaters at an elbow of capture. The "pirate" stream's flow increases, and the captured stream's flow dwindles. The victim stream may then rise below the elbow of capture, leaving a dry tract of valley called a wind gap.

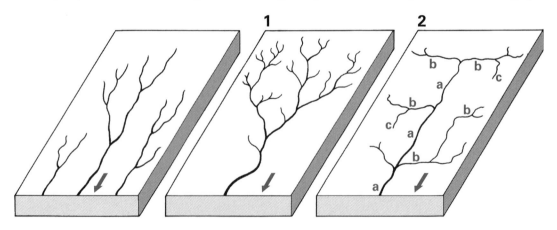

4 Accordant drainage is where river channels relate to rock type and structure. On folded rocks, tributaries quickly attack tension cracks in (upcurved) anticlinal crests and often wear these down below intervening valleys formed in (downcurved) synclines. Synclinal crests occur in the Appalachian ridge-and-valley section. (Discordant drainage patterns show no relation to today's rock structure. They may be antecedent—formed before the land was raised or tilted—or superimposed on present rocks after developing on others that have worn away.)

3 River capture
A Headward erosion of (**a**) toward (**b**)
B (**a**) has captured (**b**)'s headwaters
c Elbow of capture
d Wind gap

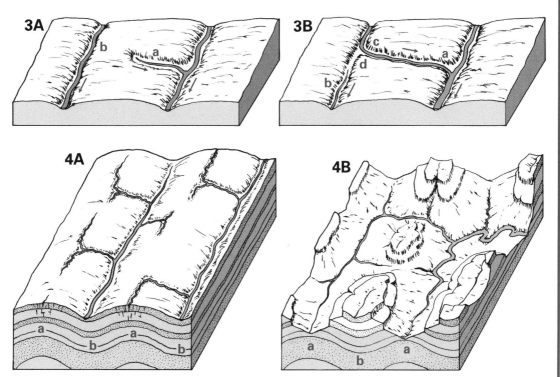

4 Accordant drainage
A Rivers flow off anticlinal slopes and along synclinal troughs
a Anticline
b Syncline
B River erosion has lowered anticlines to valleys and left synclines upstanding as mountains.

© DIAGRAM

131

7 | PLATEAUS AND RIDGES

As valleys eat into sedimentary rocks, they sculpt distinctive upland features determined largely by the type and angle of rock layers—especially where weak layers such as clay or shale alternate with more resistant layers like chalk, limestone, or sandstone.

Horizontal and tilted layers produce contrasting types of upland.

1 Plateaus develop where resistant rock caps other horizontal layers. River valleys may dissect a plateau into tablelands such as Brazil's *tableiros*, or the Colorado Plateau's steep-sided blocks called mesas, many now eroded into the smaller blocks called buttes. Plateaus may have "stepped" sides shaped by the different rates of erosion of alternating weak and resistant rock layers.

2 Cuestas are ridges formed by gently tilted strata. Each has a steep slope or escarpment and a gentle slope or dip slope. Beyond the cuesta lies lower land where erosion has eaten deeply into weaker rock. Cuestas abound in the southwestern USA, and occur along the Gulf and Atlantic coasts. Cuestas overlooking clay vales dominate southeast England's landscape and rim part of the Paris Basin.

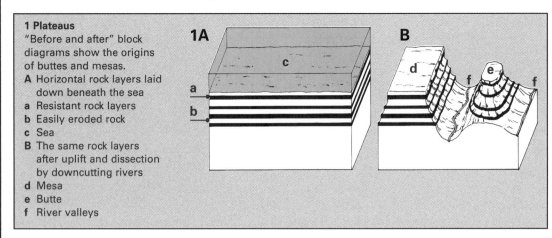

1 Plateaus
"Before and after" block diagrams show the origins of buttes and mesas.
A Horizontal rock layers laid down beneath the sea
a Resistant rock layers
b Easily eroded rock
c Sea
B The same rock layers after uplift and dissection by downcutting rivers
d Mesa
e Butte
f River valleys

3 Hogbacks are steep, even-crested ridges formed by sharply dipping strata. They abound in the mid and southern Rockies and form most of the long parallel ridges of the Appalachian Mountains' ridge-and-valley system.

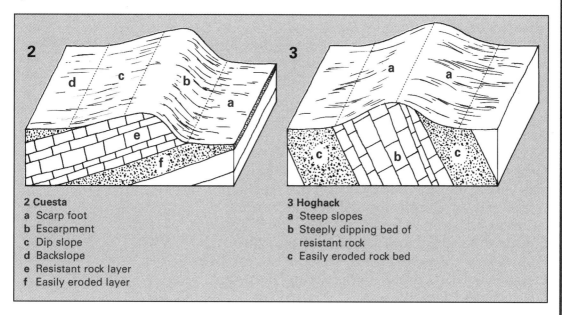

2 Cuesta
a Scarp foot
b Escarpment
c Dip slope
d Backslope
e Resistant rock layer
f Easily eroded layer

3 Hoghack
a Steep slopes
b Steeply dipping bed of resistant rock
c Easily eroded rock bed

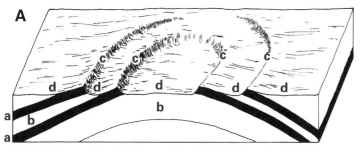

Cuestas and vales
Block diagrams show the dissection of a dome and basin of sedimentary rock layers into cuestas with intervening vales.
A Dissected dome (dip slopes facing outwards)
B Dissected basin (dip slopes facing inward)
a Resistant rock layers
b Easily eroded rock layers
c Cuestas
d Vales

© DIAGRAM

133

7 | HOW LAKES FORM

Lakes are bodies of water lying in depressions on land. They can be large or small, deep or shallow, fresh or salt. The largest lake is the Caspian Sea in central Asia; the deepest, Lake Baikal in eastern Siberia.

Lakes are formed by earth movements, volcanoes, erosion, deposition, or erosion and deposition. Here are examples of lakes formed by each.

1 Earth movements Crustal uplift isolated the Caspian Sea from the Black Sea. Crustal warping (and later glacial action) formed Lake Superior. Fault blocks sinking between high valley walls produced the long, narrow trench containing Lake Malawi and other African Rift Valley lakes.

2 Volcanic action Water-filled volcanic craters include Oregon's Crater Lake. Lava flows damming river valleys held back the waters of lakes like the Sea of Galilee and Lac d'Aydat in south-central France.

3 Erosion Lakes fill ice-worn hollows, especially in Canada and Finland. Water dissolving limestone riverbanks broadened Ireland's Shannon River to create Lough Derg. Wind eroding rock to below the water table exposes lakes in deserts.

4 Deposition Rock slides damming rivers form lakes such as Montana's Earthquake Lake. Ice dams trap lakes against glaciers and ice sheets. Oxbow lakes form where a river cuts through the necks of meanders and leaves them isolated. River sediments help trap delta lakes. Bars and dunes pond back brackish coastal lakes.

5 Erosion and deposition Glacial erosion and glacial deposits called moraines form countless lakes. Moraine-dammed cirques (ice-eroded hollows) hold circular mountain lakes called tarns. End moraines at the mouths of glaciated valleys dam long narrow lakes such as New York's Finger Lakes, England's Lake District lakes, Sweden's Glint line lakes, and Italy's Lakes Como, Garda, and Maggiore.

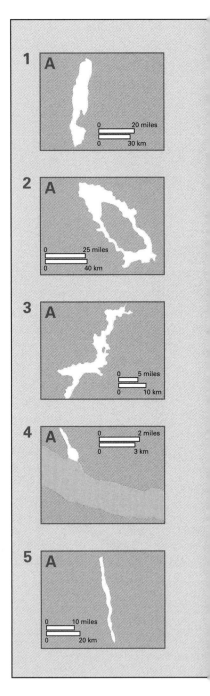

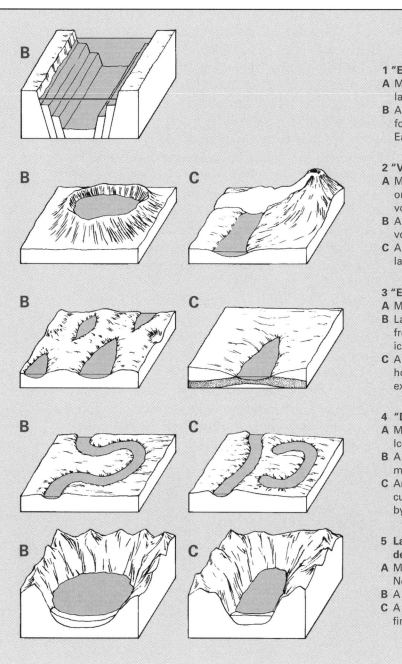

1 "Earth-movement" lakes
A Map of the Dead Sea, a salt lake in a rift valley
B A deep rift-valley lake, formed in a crack in Earth's crust

2 "Volcanic" lakes
A Map of Lake Toba, Sumatra, one of the world's largest volcanic crater lakes
B A lake occupying a volcanic crater
C A lake dammed by a lava flow

3 "Erosion" lakes
A Map of Lough Derg, Ireland
B Lakes in rock basins gouged from the rock by passing ice sheets
C A lake in a deflation hollow where desert wind exposed the water table

4 "Deposition" lakes
A Map of Lake Vatnsdalur, Iceland, dammed by ice
B A river about to cut through a meander
C An oxbow lake—a meander cut off from the river by sediment

5 Lakes formed by erosion and deposition
A Map of Lake Seneca, one of New York's Finger Lakes
B A moraine-dammed cirque
C A moraine-dammed finger lake

VANISHING LAKES

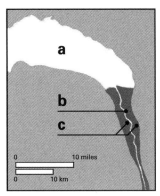

Lake Geneva *(above)*
A map shows how river sediment has filled in one end of this lake.
a Lake Geneva
b Rhône River
c Former lake, now floored by river sediment

Most lakes are geologically short-lived; they dry up in only a few thousand years or so. A lake can disappear in several ways. Many lakes get clogged by mud and silt washed in by rivers. The Rhône River will fill Lake Geneva with mud in 40,000 years.

Huge prehistoric lakes drained away or shrank with the melting of ice sheets that had ponded back their waters. Once larger than all of today's Great Lakes combined, Canada's Lake Agassiz has been reduced to the remnant Lakes Winnipeg, Winnipegosis, and Manitoba. The Great Lakes themselves are shrunken relics of mighty prehistoric Lake Algonquin. And about 15,000 years ago, the bursting of an ice dam abruptly emptied Montana's mighty prehistoric Lake Missoula in a flood some ten times greater than the flow of all the rivers in the world.

Some lakes dry up because the local rainfall dwindles and the lakes lose more water by evaporation than they gain from rivers. A drying climate helped shrink the Great Basin lakes of the United States.

Outlet rivers cutting down through bedrock are another factor. Utah's Great Salt Lake covers one-tenth the area of its precursor, Lake Bonneville, which lost one-third its volume in about six weeks through a breach at Red Rock Canyon.

Lost lakes of North America
(right)
Maps show lakes lost in relatively recent times, with their major modern relics.
A North-central North America
a Former Lake Agassiz
b Lake Winnipeg
c Lake Winnipegosis
d Lake Manitoba
e Lake of the Woods
B Great Basin lakes
f Former Lake Lahontan
g Former Lake Bonneville
h Great Salt Lake

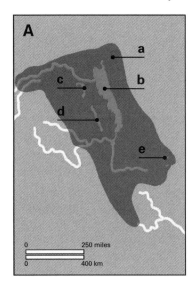

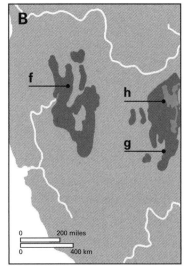

Disappearing lakes and ponds produce many of the world's wetlands. Swamps are places that are always waterlogged; for instance, the Florida Everglades, Virginia's Dismal Swamp, and much of the Sudd in the Sudan. Marshes are low-lying lands that flood when rivers overflow. Bogs are soft, wet, spongy areas where moss fills shallow lakes and pools.

When even wetlands dry out, clues to vanished lakes remain in flat valley floors thickly carpeted with mud and silt, old overflow channels, and old beaches high on valley slopes.

A lake vanishes (below)
Diagrams show how a river fills in a lake.
A River washes sediment into one end of the lake.
B River sediment builds a delta out into the lake.
C Sediment shrinks the lake into a small, shallow, reedy swamp.
D The lake floor is now filled in with sediment colonized by land plants.

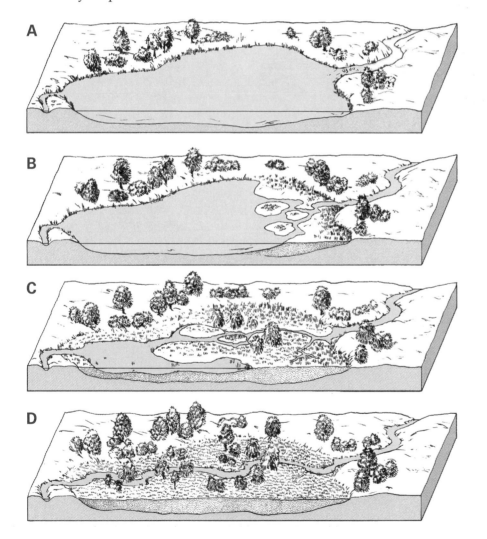

A

B

C

D

CHAPTER 8 THE WORK OF THE SEA

Seawater set in violent motion scours coasts around the world. On rocky shores, waves armed with stones batter the land away. Yet fragments torn from cliff-fringed coasts lodge on shores to raise new sand and shingle beaches. This chapter shows how the sea erodes and builds the land, and describes shores risen or submerged by the changing levels of the sea or land. It ends with a discussion of the growth of coral reefs and islands.

A nineteenth-century engraving showing the dramatic rock formations of the Scottish coast.

8 THE SEA IN ACTION

Besides shaping inland surfaces, water sculpts the coast—the zone where land meets sea. Coasts include sea cliffs, shores (areas between low water and the highest storm waves), and beaches (shore deposits). Seawater set in motion erodes cliffs, transports eroded debris along shores, and dumps it on beaches. Therefore, most coasts retreat or advance. The chief agents in this work are waves and currents, but tides contribute too.

1 Waves are undulations set in motion mainly by the wind. Wave height and power depend upon wind strength and fetch—the amount of unobstructed ocean over which the wind has blown. In the open sea, waves pass through the water without moving it forward. But in shallow water, wave crests crowd closer, pile up, and overbalance, causing forward movement of the water. The consequences are featured later in this chapter.

2 Underwater currents called undertows flow away from the shore, balancing the onshore pileup of water by waves. Rip currents are strong local currents of this kind.

3 Tides are two sets of huge progressive waves that sweep around the oceans each day. They are caused by the gravitational pull of the Moon and Sun. Earth's spin, continents, coastlines, and underwater ridges affect local tidal height. Coinciding with storm waves, the highest tides affect the highest level of the shore.

1 Waves *(below)*
Waves trigger the circling of water particles in a stack about half a wavelength deep. In shallower water the circles are flattened by their proximity to the bottom. The waves grow in height, and become unstable, and their crests topple and break.
a Wave crest
b Wave trough
c Wave height
d Wave length

1

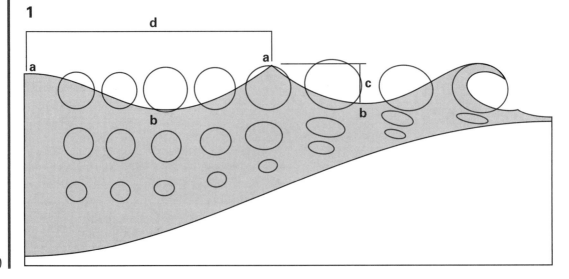

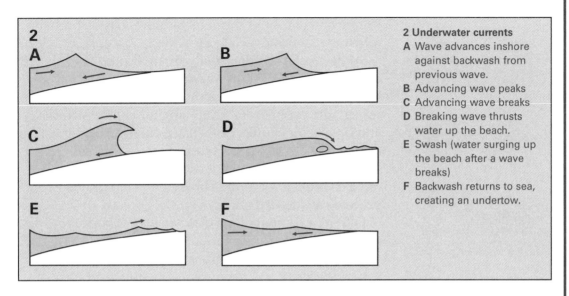

2 Underwater currents
A Wave advances inshore against backwash from previous wave.
B Advancing wave peaks
C Advancing wave breaks
D Breaking wave thrusts water up the beach.
E Swash (water surging up the beach after a wave breaks)
F Backwash returns to sea, creating an undertow.

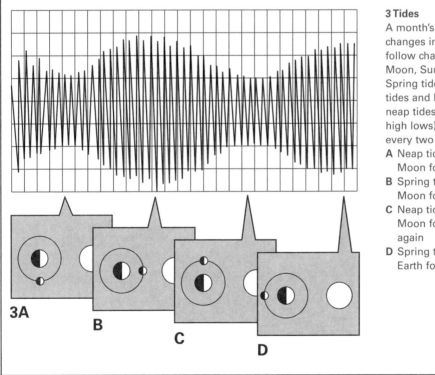

3 Tides
A month's tides show how changes in tidal range follow changes in Earth, Moon, Sun alignment. Spring tides (high high tides and low low tides) and neap tides (low highs and high lows) each occur about every two weeks.
A Neap tide: Sun, Earth, Moon form a right angle
B Spring tide: Sun, Earth, Moon form a straight line
C Neap tide: Sun, Earth, Moon form a right angle again
D Spring tide: Sun, Moon, Earth form a straight line

8 SEA ATTACKS THE LAND

Where cliffs or rocks rim the land, the coast is probably retreating as waves erode the shore.

Wave erosion works chiefly by hydraulic action, corrasion, and attrition. As waves strike a sea cliff, hydraulic action crams air into rock crevices; as waves retreat, the explosively expanding air enlarges cracks and breaks off chunks of rock. Chunks hurled by waves against the cliff break off more pieces—a process called corrasion. Rubbing against each other and the cliff reduces broken rocks to pebbles and sand grains—a process called attrition.

Different combinations of wave action, rock type, and rock beds produce these features:

1 Cliff and wave-cut platform A sea cliff forms where waves undercut a slope until its unsupported top collapses. As waves eat farther back inland, they leave a wave-cut bench or platform jutting out beyond the cliff, below the sea.

1 Cliff and wave-cut platform
a Sea cliff being formed by undercutting wave erosion
b Wave-cut platform
c Beach deposits
d High tide level
e Low tide level

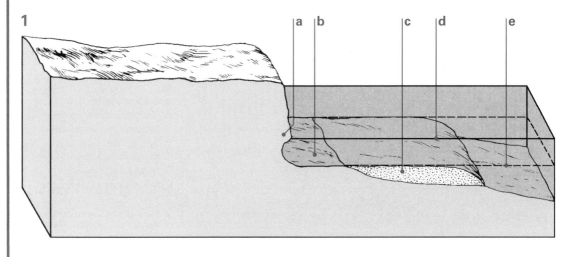

2 Cliff steepness
A Gentle cliff slope influenced by landward-tilted rock layers
B Steep cliff influenced by seaward-tilted rock layers

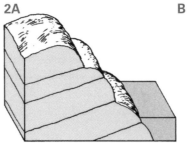

2 Cliff steepness varies with rock hardness (usually the harder the rock, the steeper the cliff) and the angle of rock layers. Landward-tilted layers tend to produce a gentle cliff slope; seaward-tilted layers may give an overhanging cliff.

3 Headlands and bays Resistant rock juts out as a headland after erosion of nearby less resistant rock has eaten out a bay. Subsequent erosion produces caves, blowholes, arches, and stacks. At a sea cliff base, wave action may enlarge a horizontal crack to gnaw out a sea cave. Inside, erosion of a vertical joint may form a clifftop blowhole. A sea cave driven through a headland forms an arch. Roof collapse isolates the headland's tip as a steep-sided island called a stack.

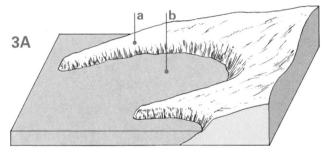

3A

3 Headlands and bays
Three stages in cliff erosion are:
A Formation of headland (**a**) and bay (**b**)
B Erosion cuts sea caves (**c**) and a clifftop blowhole (**d**)
C Cave erosion creates arches (**e**) and, later, stacks (**f**)

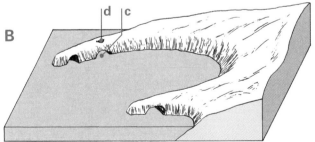

B

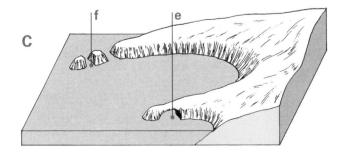

C

DROWNED COASTLINES

Drowned coastlines
Maps (**A**) and block diagrams
(**B**) show corresponding types
of drowned coastline.
1A Dalmatian Croatia
1B Dalmatian coast
2A Southern Alaska
2B Fjord coast
3A Southwest Ireland
3B Ria coast
4A Thames rivermouth
4B Estuary coast
5A Southern Sweden
5B Fjard coast
6A Boston Harbor
6B Coast with submerged
    glacial deposits

If land subsides or sea level rises, the sea invades low-lying areas, drowning coastal plains, invading valleys, converting ridges to peninsulas, and isolating uplands as islands. Submerged upland coasts include the Dalmatian coastlines, fjords, and rias. Submerged lowland coasts may feature estuaries, fjards, and submerged glacial deposits.

1 Dalmation coastlines (also known as drowned concordant, longitudinal, or Pacific coastlines) feature mountain ridges that parallel the sea. Flooding formed valleys into sounds, and isolated ridges as long, narrow offshore islands. Croatia's Dalmatian coast and many British Columbian islands were shaped this way.

2 Fjords are submerged, glacially deepened inlets with sheer, high sides, a U-shaped cross profile, and a submerged seaward sill largely formed of end moraine. Fjords occur in south Alaska, British Columbia, south Chile, Greenland, New Zealand's South Island, and Norway.

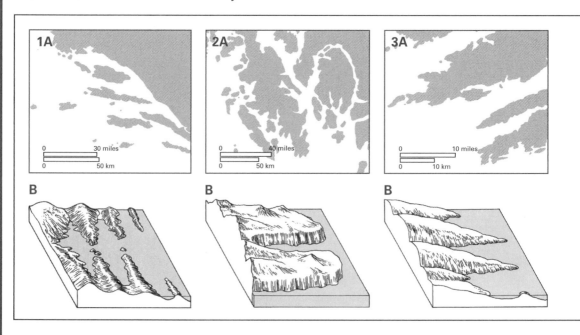

3 Rias are drowned river valleys forming long, funnel-shaped, branching inlets meeting the sea at right angles and with a V-shaped cross profile. Rias abound in southwest England, southwest Ireland, northwest France, and northwest Spain.

4 Estuaries are tidal river mouths, many of them drowned low-lying river valleys, flanked by mudflats and pierced by a maze of creeks submerged at high tide. Such estuaries indude those of the Elbe, Gironde, Hudson, St. Lawrence, Chesapeake (Susquehanna), and Thames.

5 Fjards are ice-deepened lowland inlets often with small islands at the seaward end. They indent rocky, glaciated lowlands, as in southern Sweden, Nova Scotia, and the Shetland Islands.

6 Submergence of glacial deposits has left drumlins as offshore islands in Boston Harbor and Northern Ireland's Strangford Lough.

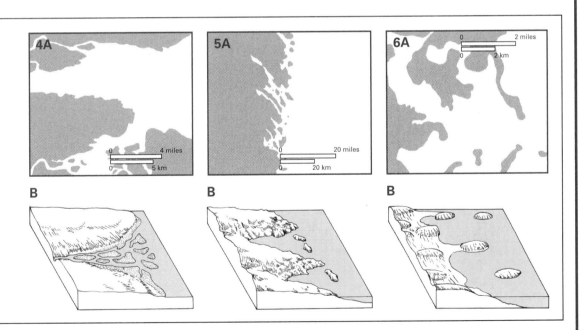

© DIAGRAM

8 HOW SEA BUILDS LAND

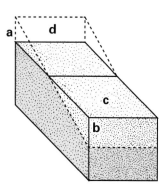

Shore deposition
On a gently sloping shore (a–b) sea erodes the lower part (c) and dumps it on the upper part (d). On a steeply sloping shore the opposite occurs. Most beaches are in near equilibrium.

Rock fragments torn from one stretch of coast are often added to another. Storm waves may hurl seabed sand and shingle inshore. Waves breaking at an angle to the land produce the longshore (littoral) drift of sand and shingle along the beach. Where land slopes gently to the waves, extensive beaches form, and land may grow into the sea. But storms shift or strip away huge quantities of loose material, especially on steeply sloping upper beaches. In the end, gravity transfers most beach material to the seabed.

Coasts of deposition include these items:
1 Boulder beaches are narrow belts of rocks and shingle at the base of sea cliffs.
2 Bay-head beaches are small sandy crescents. Each lies in a cove between two rocky headlands.
3 Lowland beaches are broad, gently sloping sandy beaches with a strip of shingle on the upper shore, often backed by dunes of sand blown inland by onshore winds. Such shores rim the southern Baltic and form the Landes of southwest France.
4 Bar A bar is an offshore strip of sand or shingle parallel to the coast. Bars border most of the US Southeast and Gulf coastline. They include the barrier islands that form Cape Hatteras.

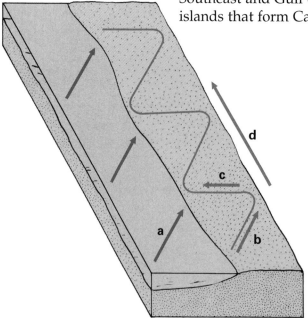

Longshore drift
a Wind and wave direction
b Direction of swash
c Backwash
d Direction of drift

5 Spit A spit resembles a bar, with one end tethered to the land. Special forms include bay-bars that link two headlands, tombolos linking islands to a mainland as in England's Chesil Beach, and (some) cuspate forelands—triangular shingle formations such as Cape Canaveral and Dungeness, England.

6 Mud-flats and **salt-marshes** form in estuaries and sheltered bays. Mud-flats are barren beds of silt and clay deposited by tides and drowned at high tide. Silt-trapping plants raise their level, and create salt marsh drained by tidal creeks. In the tropics, mud-flats colonized by mangrove trees form mangrove swamps.

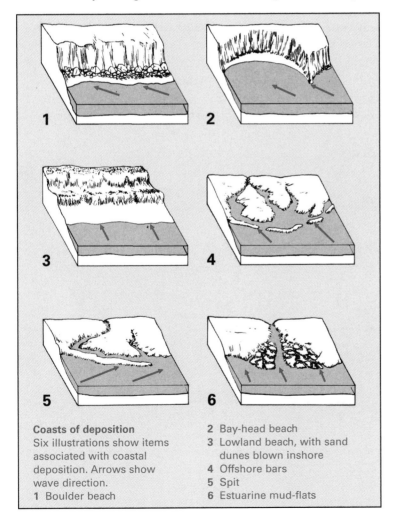

Coasts of deposition
Six illustrations show items associated with coastal deposition. Arrows show wave direction.
1 Boulder beach
2 Bay-head beach
3 Lowland beach, with sand dunes blown inshore
4 Offshore bars
5 Spit
6 Estuarine mud-flats

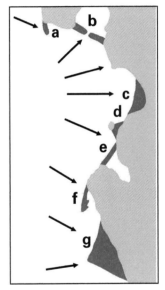

Depositional features
a Spit
b Double spits
c Bay-head beach
d Tombolo
e Barrier beach
f Hooked spit
g Cuspate foreland

© DIAGRAM

8 | SHORES RISEN FROM THE SEA

1 Emergent highland coast
a High tide level
b Low tide level
A Before emergence
c Sea cliff
d Sea caves
e Beach
f Wave-cut platform
B After emergence
g Old sea cliff
h Old sea caves
i Raised beach
j Exposed wave-cut platform
k New sea cliffs
l New beach

While sinking land or rising sea level has drowned some coasts, rising land or falling sea level has stranded other coasts above the present level of the waves. Raised highland and lowland coasts have somewhat different characteristics.

1 Emergent highland coasts feature raised beaches. A raised beach, often covered with shells or shingle, stands perched high and dry above sea level. The raised beach is an old shoreline and adjoining wave-cut rock platform, up to 2,600 feet (about 800 m) across. Inland, above the raised beach, rises an old sea cliff, perhaps pierced by wave-cut caves. Below the raised beach, a new sea cliff and wave-cut platform form the seaward boundary. Successive earth movements can create a series of raised beaches, one above the other. Many formed where land depressed by ice bobbed up once the ice melted. Such beaches occur on North

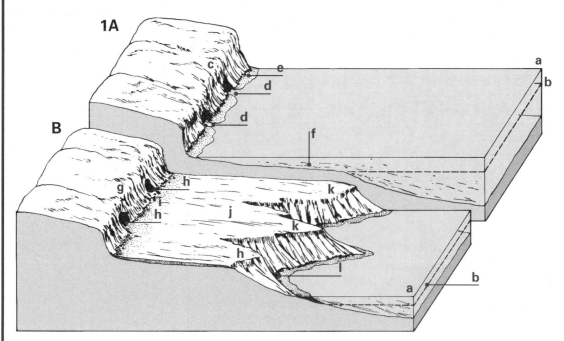

America's Arctic coasts, in western Scotland, and around the Baltic Sea. Others crop up as far apart as Malta and the South Pacific.

2 Emergent lowland coasts are coastal plains that slope gently to the sea and are rimmed by marshes, sandy beaches, bars, lagoons, and spits. The plains are uplifted continuations of the shallow offshore continental shelf, and thus may be floored by seashells, sand, and clay consolidated into limestone, sandstone, and shale. Inland, the plains may end abruptly below a line of hills marking the old coastline. This happens in the southeast United States—its coastal plain abuts the fall line where rejuvenated rivers flow steeply from the Appalachian Mountains to create a line of waterfalls. Other emergent lowland coasts include the northern Gulf of Mexico, the southern Rio de La Plata, and much of the east coast of India.

2 Emergent lowland coast
A Before emergence
a Gentle slopes
b River valleys
c Coastline
d Continental shelf
B After emergence
e Valleys deepened by rejuvenated rivers
f Old coastline
g New coastal plain
h New coastline

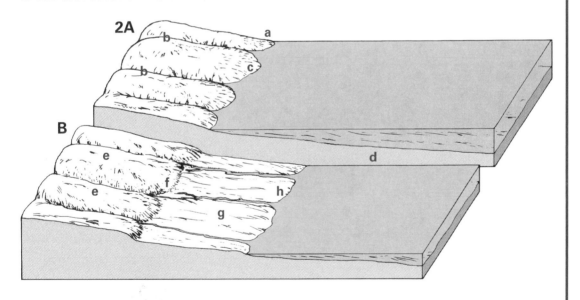

8 WHERE CORAL GROWS

Coral reefs rim shores and form low islands mainly in the world's warm seas and oceans. Coral is a limestone rock produced by tube-shaped skeletons of billions of coral polyps—animals resembling tiny sea anemones—and by limy algal plants called nullipores. Reefs grow up and out as new organisms build on the old skeletons. Waves pounding the seaward edge hurl broken chunks of coral on the reefs. Meanwhile sand accumulates upon the shoreward side, and reefs become low islands.

Coral polyps need clear, warm, shallow, salty water, found mostly in tropical seas and oceans to the east of continents. But many prehistoric reefs now lie inland in rocks outside the tropics.

There are three main types of coral reef.

1 Fringing reefs are narrow offshore coral reefs, separated by a shallow lagoon from the nearby coast. Such reefs abound off the Bahamas and Caribbean islands.

1 Fringing reef
a Edge of land
b Narrow, shallow lagoon
c Living coral
d Mound of dead coral
e High water
f Low water

2 Barrier reef
a Edge of land
b Wide, deep lagoon
c Living coral
d Mound of dead coral
e High water
f Low water

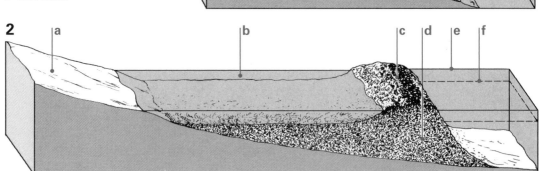

3A

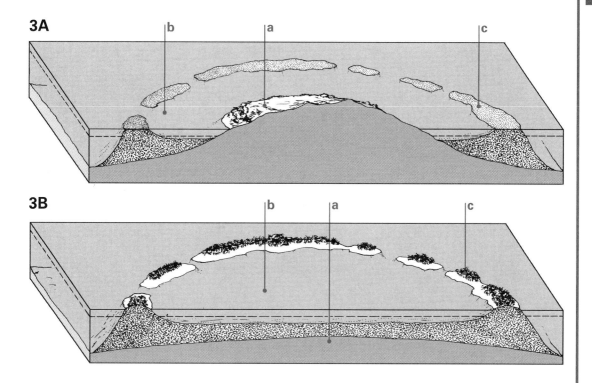

3B

2 **Barrier reefs** are broad coral platforms separated from the coast by a wide, deep channel. At least some formed upon subsiding coasts. The world's largest is northeast Australia's Great Barrier Reef, 1,260 miles (2,027 km) long.

3 **Atolls** comprise circular chains of coral reefs. Each atoll encloses a lagoon and probably started as a fringing reef around a volcanic island. As the island began to sink and/or sea level rose, coral growth kept pace. Thus the reef became a barrier reef and then an atoll. Borings show that Eniwetok Atoll in the Pacific Ocean grew upward from a volcanic base now more than 4,600 feet (1,400 m) below the surface. Atolls abound chiefly in parts of the Pacific and Indian Oceans. The Pacific's Marshall Islands include the world's largest atoll, Kwajalein—more than 170 miles (274 km) long.

3A Atoll forming
a Slowly sinking volcanic island
b Lagoon
c Barrier reef

3B Atoll formed
a Submerged volcanic island
b Lagoon
c Coral reef

© DIAGRAM

CHAPTER 9 THE WORK OF ICE AND AIR

Ice corrodes mountains into sharp-tipped peaks and saw-edged ridges. Moving rivers of ice deepen valleys and bevel hills. But melting ice sheds sheets and strips of debris on lowlands, and near an ice sheet, freezing and thawing rework the ground.

Where wind scours deserts, blown sand sculpts rocks and wears hollows in the land. But windblown particles accumulate as dunes and sheets. The work of wind and water creates distinctive landscapes.

Three panoramic views of the Swiss Alps. (Engravings taken from an old edition of *Baedeker's Guide to Switzerland*)

153

GLACIERS AND ICE SHEETS

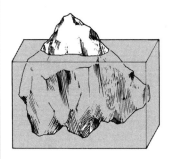

Castellated berg
These bergs plunge into the sea from Arctic glaciers. Vast flat-topped bergs break off the ice shelves that fringe Antarctica. Polar seas also bear vast tracts of sea ice.

Ice covers about ten percent of all land and 12 percent of the oceans. Most lies in polar sea ice, polar ice sheets and ice caps, valley glaciers, and piedmont glaciers formed by valley glaciers merging on a plain.

Most of Antarctica lies beneath an ice sheet twice the size of Australia and up to 14,000 feet (4,300 m) thick. Another ice sheet covers Greenland. Such vast slabs of ice spread slowly, smothering all but a few projecting peaks, or nunataks. Smaller ice masses, called ice caps, crown parts of Iceland, Norway, and some Arctic islands.

Valley glaciers are tongues of ice in mountain ranges. They start in ice-worn rock basins called cirques. Here, old snow forms firn or névé—a mass of ice pellets compacted by the weight of snow above. Fed by fresh

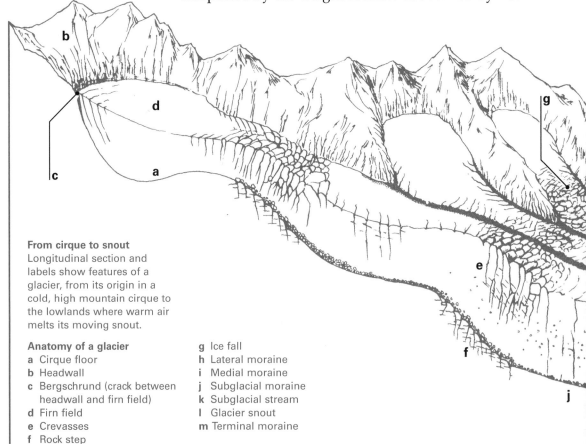

From cirque to snout
Longitudinal section and labels show features of a glacier, from its origin in a cold, high mountain cirque to the lowlands where warm air melts its moving snout.

Anatomy of a glacier

a Cirque floor
b Headwall
c Bergschrund (crack between headwall and firn field)
d Firn field
e Crevasses
f Rock step

g Ice fall
h Lateral moraine
i Medial moraine
j Subglacial moraine
k Subglacial stream
l Glacier snout
m Terminal moraine

snowfalls, firn spawns a glacier that spills down a valley filling it with ice, perhaps for scores of miles.

Glaciers creep downhill at an inch (2.5 cm) to 100 feet (30 m) a day, depending on conditions. Pressure makes the lower ice plastic enough to flow, but the upper ice stays rigid. Varying gradients and differential rates of flow within the glacier split the surface with deep cracks called crevasses. Where a glacier plunges down a steep rock step, the surface forms an ice fall—a crisscross maze of crevasses and isolated pinnacles.

Frost-shattered rocks and stones falling on the glacier's flanks form lines of lateral moraines. Where a tributary glacier joins a major glacier, two lateral moraines merge as a medial moraine.

Most valley glaciers end in a moraine-rich snout where warm air melts ice as fast as the glacier flows. If ice melts any faster, the snout retreats.

In polar regions, glaciers and ice sheets extend down to the sea. Huge chunks of ice snap off and float away as icebergs. Antarctica's Ross Barrier is an example of a floating slab of ice as big as Spain that remains tethered to the land.

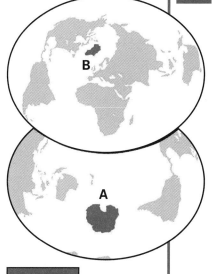

Frozen landmasses
Maps locate two landmasses mostly under ice.
A Antarctica
B Greenland
Block diagrams compare their ice sheets with the areas of two other countries.
C United States
D Mexico

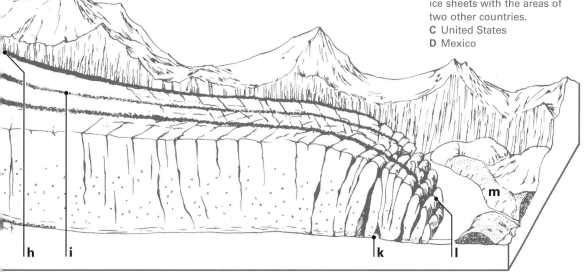

© DIAGRAM

9 | HOW ICE ATTACKS THE LAND

A

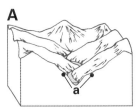

B

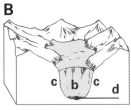

C

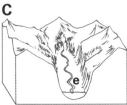

Gnawed by ice (above)
Three diagrams show the major changes wrought by glaciation in a mountain valley.
A Before glaciation
a V-shaped cross section
B During glaciation
b Glacier
c Ice-scoured valley sides
d Ice-deepened valley floor
C After glaciation
e Glacial trough with a U-shaped cross section

Ice-worn hummocks (right)
A Roche moutonnée
a Gently sloping upstream side, grooved by stones in moving ice
b Steep, rough, ice-plucked downstream side
B Crag-and-tail
c Ice-rubbed resistant crag
d Protected tail of soft rock

When glaciers and ice sheets melt, they leave a landscape scraped by moving ice. Ice sheets armed with broken rocks beveled huge tracts of northern Canada and Finland. Frost and moving ice have sharpened peaks and deepened mountain valleys in the Andes, Alps, Himalayas, and Rockies.

Valley-glacier erosion starts high up on hills or mountainsides. Here, freeze-thaw processes fracture rocks, as well as enlarge and deepen shallow snow-filled dips, to form the steep rock basins known as corries, cwms, or cirques. Two cirques eating back into a mountain may leave a knife-edged ridge called an arête, and sometimes the cirques make a notched, narrow pass called a col. Where at least three cirques converge, arêtes meet in a pyramidal peak. The famous Matterhorn originated in this way.

A glacier that fills a mountain valley shoves loose material ahead and plucks rocks from the valley sides. Loosened stones embedded in the frozen river's flanks scrape more rock from the valley sides. Thus a glacier widens, deepens, and straightens; it changes a valley's V-shaped cross section into the U-shape of the glacial trough. (Glacial troughs invaded by the sea became the fjords of Alaska, Norway, and New Zealand.)

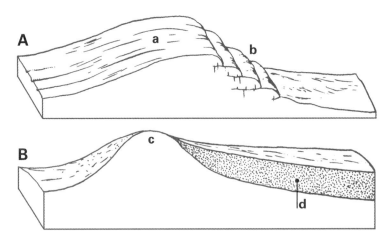

Glacial erosion truncates (lops off) the spurs that jut out into the main valley, and leaves tributary streams to plunge from the lips of the hanging valleys that end high above the bottom of the trough. Yosemite's Bridalveil Fall is such a waterfall. Rock debris embedded in moving ice scratches and polishes the valley walls and floor, producing telltale grooves. Similar striations betray the ice sheet that once covered New York's Central Park. Other telltale features include two distinctive types of ice-worn hummocks known as crag-and-tail and roche moutonnée.

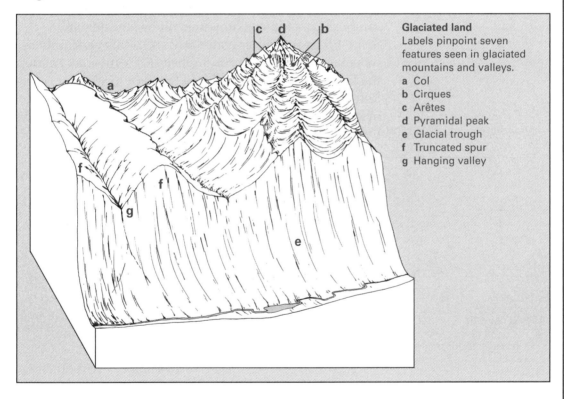

Glaciated land
Labels pinpoint seven features seen in glaciated mountains and valleys.
a Col
b Cirques
c Arêtes
d Pyramidal peak
e Glacial trough
f Truncated spur
g Hanging valley

9 | DEBRIS DUMPED BY ICE 1

Glaciers and ice sheets transport in, on, and beneath the ice from where they are formed, substances from finely powdered rock to mighty boulders. Streams emitted by the ice transport more ice-eroded debris. Where glaciers and ice sheets melt, this vast load of drift material remains and modifies the land. Experts calculate that more than one-third of Europe, nearly one-quarter of North America, and one-eighth of the world's land surface is cloaked in debris shed by ice or meltwater.

Debris dumped by glaciers themselves is an unsorted mass of stones and rock embedded in a sandy, clayey matrix known as till or boulder clay. Some till forms under active ice, and some accumulates where ice decays. How and where till forms determines the landscape features in huge tracts of lowland. Retreating ice left the undulating sheets called ground moraine that cover much of the North European plain. Drift added flat floors to many U-shaped glaciated valleys. And where an ice front paused, you will find the long, curved ridges of its terminal or end moraines.

Drumlins are elongated hummocks up to 300 feet (90 m) high and l mile (1.6 km) long. Many formed where valley glaciers shed and streamlined drift as they reached a plain and spread out.

Erratic blocks had other origins. Many of these ice-borne rocks lie far from where they started. Some Scottish rocks have ended up in southeast Ireland, and 600 miles (1,000 km) separate Kentucky's red jasper boulders from their nearest possible source north of Lake Huron. Erratic blocks can be enormous. One Albertan specimen reportedly exceeds 18,000 tons.

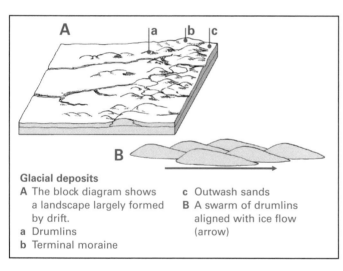

Glacial deposits
A The block diagram shows a landscape largely formed by drift.
a Drumlins
b Terminal moraine
c Outwash sands
B A swarm of drumlins aligned with ice flow (arrow)

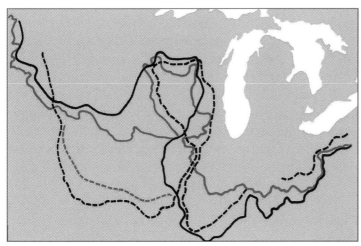

Midwest drift deposits
This map marks the southern limits of the glacial deposits laid down by melting ice sheets in the Midwest. Each line marks the farthest advance of a different glacial stage, from Nebraskan (early) to Wisconsin (late).

〜〜〜 Wisconsin

〜〜〜 Illinoian

▬ ▬ ▬ Kansan

▬ ▬ ▬ Nebraskan

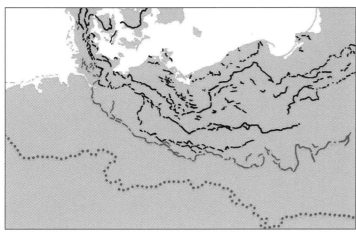

North European moraines
Here we show the southern limits of the glacial deposits laid down by successive, melting ice sheets based on the landscape of Scandinavia. Terminal moraines formed where a melting ice front paused.

〜〜〜 Recent terminal moraines

‑〜‑〜 Earlier terminal moraines

• • • • • Southern edge of glaciation

Erratic block
An immense granite boulder taller than a man lies on limestone. A melting ice sheet dumped the boulder more than 10,000 years ago.

© DIAGRAM

DEBRIS DUMPED BY ICE 2

Meltwater streams that issue from a glacier or ice sheet produce layered sediments; these therefore differ from the unsorted drift directly dropped by ice. Geologists call them outwash or glaciofluvial deposits.

Outwash takes two main forms: outwash plains and valley trains. Outwash plains are layered sheets of clay, sands, and gravels that fan out over lowlands ahead of where an ice sheet lay. The largest particles settle near the ice rim. Finer particles may travel many miles. Valley trains are thick outwash deposits covering the floors of deep, narrow, glaciated valleys.

Outwash deposits give rise to the following features:
1 Eskers These long, winding ridges are aligned with the flow of retreating glaciers or ice sheets. They are from a few feet to a few hundred feet wide, up to 100 feet (30 m) high and some extend for many miles. They grow where subglacial streams shed their loads in ice tunnels or at the tunnels' receding mouth. Eskers are plentiful in lowland areas around the Baltic Sea.

Glacial valley
a Glacier
b Blocks of melting ice
c Lake filled by sediment
d Surface streams
e Crevasses containing sediment shed by streams
f Ice-margin lake
g Deltas
h Subglacial stream
i Englacial stream

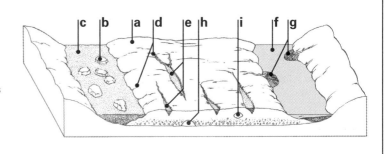

Postglacial valley
a Exposed valley floor
b Kettles (ex ice-filled hollows)
c Kame-terrace (old lake bed)
d Streams
e Kames (ex crevasse deposits)
f Kame-terrace (old lake bed)
g Kame-deltas
h Esker from subglacial stream
i Esker from englacial stream

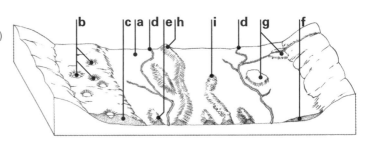

2 Kames are mounds that come in two main forms called kame-terraces and kame-moraines. A kame-terrace is a narrow, flat-topped, steep-sided ridge of sediments along a valley side. It formed below a stream or lake trapped between the valley side and a prehistoric glacier. Kame-moraines, or kame-deltas, are complex undulating mounds of sands and gravels dumped along a stagnant ice sheet's rim. Kame-moraines are plentiful in central Ireland and the United States between Long Island and Wisconsin.

3 Kettles are hollows in kame-terraces or kame-moraines, formed where ice blocks have melted. Such ice blocks occur today in Iceland.

4 Varve clays are banded layers of fine and coarse material deposited in meltwater lakes fringing ice sheets. Coarse material washed in with the summer thaw. Fine material settled in winter when the lakes froze over. Counting the bands enables geologists to measure post-glacial time year by year in some areas of North America and Sweden.

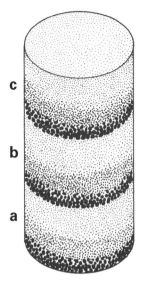

Varve clays *(above)*
This core sample shows three banded layers, representing sediments laid down in three successive years in a lake liable to winter freezing.
a First year
b Second year
c Third year

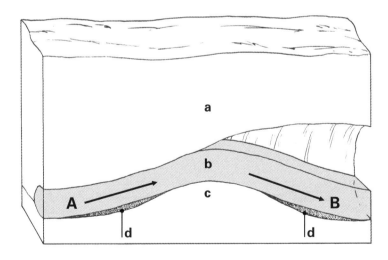

Subglacial sediments *(left)*
Here we show two situations where a stream beneath a glacier may shed its load:
A The stream flows uphill and loses carrying power.
B The channel widens, so water flows at reduced pressure.
a Ice
b Stream
c Bedrock
d Sediment

© DIAGRAM

AROUND AN ICE SHEET'S RIM

Prolonged freezing and brief summer thawing around an ice sheet's rim produces the periglacial ("around the ice") or tundra landscapes of far northern Eurasia and North America. Some periglacial features dating from the last glaciation still show in lands much farther south.

A major feature is permafrost—permanently frozen ground beneath the surface. In places, Siberian permafrost extends 2,000 feet (610 m) deep. Above it lies the active layer, 6–20 feet (2–6 m) thick. In summer, meltwater fills and lubricates the active layer; sloping surfaces then creep downhill upon the frozen layer beneath. Known as solifluction, this process dumps head deposits of frost-shattered rocks, like the chalk-rubble coombe deposits still seen in parts of southern England.

Freeze-thaw affects both solid rock and loose materials. On north-facing slopes, freeze-thaw beneath

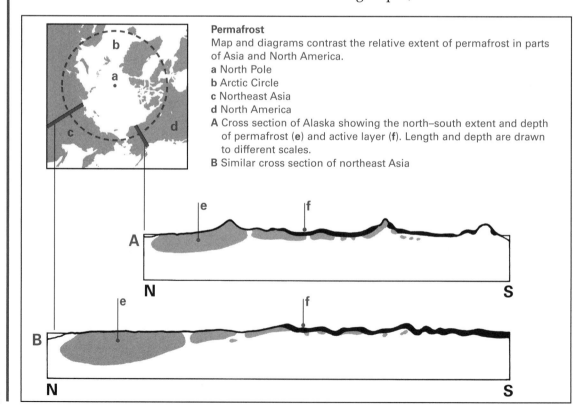

Permafrost
Map and diagrams contrast the relative extent of permafrost in parts of Asia and North America.
a North Pole
b Arctic Circle
c Northeast Asia
d North America
A Cross section of Alaska showing the north–south extent and depth of permafrost (**e**) and active layer (**f**). Length and depth are drawn to different scales.
B Similar cross section of northeast Asia

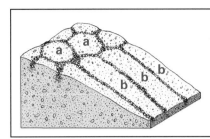

Patterned ground
a Stone polygons one yard (1 m) or more across, with sorted fine material inside
b Stone stripes: parallel lines of stones formed on steep slopes under the influence of soil creep

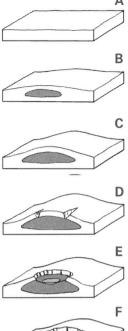

snow patches loosens bits of bedrock, and meltwater and solifluction carry them away. Called nivation, this sequence wears big nivation hollows into the rock. Below loose surfaces, freezing water expands, so the ground heaves. Repeated heaving sorts out large and small soil particles. The result is patterned ground with stones arranged in circles, nets, polygons, and stripes.

Other hallmarks of the periglacial fringe include ice wedges, thermokarst, and pingos.

Ice wedges form where cracks appear in frozen ground. Meltwater fills the cracks and freezes as ice wedges that taper downward for as much as 35 feet (10 m). Gravel often fills old ice wedges.

Where ground ice melts it often leaves a surface pocked by hollows, superficially like a karst limestone surface, and thus called thermokarst.

The strangest periglacial features are pingos—cones or domes 20–300 feet (6–90 m) high. These earth or gravel humps occur where freezing water expands between an almost frozen surface and the permafrost, pushing up an earth or gravel blister.

Pingo formation (above)
A Prepingo land surface
B Ice lens forms underground.
C Expanding ice lens pushes up a dome in the land surface.
D Tension cracks in the dome expose the ice lens.
E Melting ice creates a pond.
F Sediment collects on the pond floor.

Ice

Water

Sediment

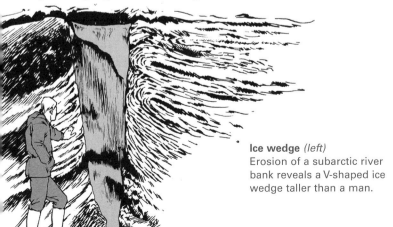

Ice wedge (left)
Erosion of a subarctic river bank reveals a V-shaped ice wedge taller than a man.

WIND THE ERODER

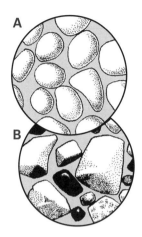

Sands compared *(above)*
A Polished, rounded,"frosted" desert grains
B More angular sand grains from a river bed

In tropical and midlatitude deserts, wind picks up specks of weathered rock and hurls them far across the barren land. Fine particles of silts and clays whirl high above the ground in dust storms. Sand grains hop and skip across the countryside. Tiny pebbles roll and slide. The process of sandblasting does the most to bevel desert rocks. Attacking pebbles, soft rocks, and rock crevices, windblown desert sand creates the erosion features shown below.

1 Ventifacts are stones with surfaces smoothed and flattened through prolonged attack by windblown sand.

2 Rock pedestals are mushroom-shaped rocks, often made of horizontal rock layers. Sand gnaws into their bases, but winds are seldom strong enough to lift sand grains above waist height. (Sand can similarly cut rock caves.)

Wind the transporter
Wind moves different particles at different levels.
a Dust particles
b Sand grains
c Tiny pebbles

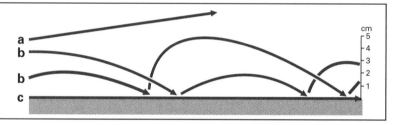

3 Zeugen are parallel, flat-topped ridges of hard rock up to 100 feet (30 m) high. They are left standing when sand has widened joints in horizontal hard rock and gnawed into the softer rock beneath.

4 Yardangs are parallel ridges of hard rock up to 50 feet (15 m) high. They form where alternating hard and soft rock layers were upended. Wind gnaws the soft rock into furrows, but leaves the hard rock standing. Yardangs occur especially in Central Asia and Chile's Atacama Desert.

5 Rock pavement, or **hamada**, is a flat, wind-smoothed rocky desert surface (not shown).

6 Deflation hollows are worn or deepened in a desert surface by the wind. Egypt's Qattara Depression—the world's largest deflation hollow—is about as big as

New Jersey, and as much as 400 feet (121 m) below the level of the sea. It is partly a tectonically formed feature, further eroded by the wind. Southwest Africa, western Australia, and Mongolia have smaller wind-worn "saucers." Some expose supplies of underground freshwater, which produce swamps or lakes, or support fertile oases.

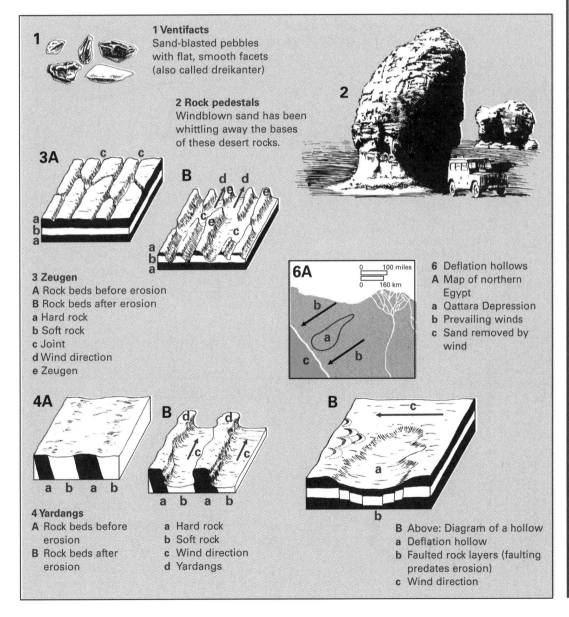

1 Ventifacts
Sand-blasted pebbles with flat, smooth facets (also called dreikanter)

2 Rock pedestals
Windblown sand has been whittling away the bases of these desert rocks.

3 Zeugen
A Rock beds before erosion
B Rock beds after erosion
a Hard rock
b Soft rock
c Joint
d Wind direction
e Zeugen

4 Yardangs
A Rock beds before erosion
B Rock beds after erosion
a Hard rock
b Soft rock
c Wind direction
d Yardangs

6 Deflation hollows
A Map of northern Egypt
a Qattara Depression
b Prevailing winds
c Sand removed by wind

B Above: Diagram of a hollow
a Deflation hollow
b Faulted rock layers (faulting predates erosion)
c Wind direction

© DIAGRAM

165

9 | WINDBLOWN DEPOSITS

In barren lands wind freely moves huge quantities of tiny particles eroded from the rocks. In deserts and near prehistoric ice sheets, these windblown fragments cloak vast areas with shifting sands or layers of cemented dust.

In some deserts winds, slowed by passing over pebbles, shed sand in smooth or undulating sheets. But winds blowing steadily from one direction pile sand into the mobile hillocks known as dunes. Those described below are the best-known forms.

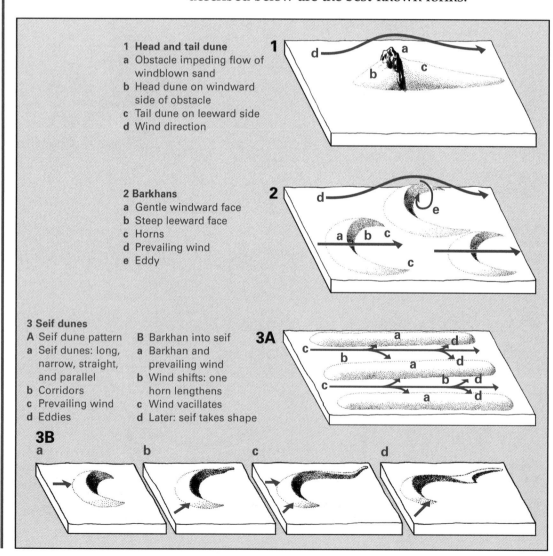

1 Head and tail dune
a Obstacle impeding flow of windblown sand
b Head dune on windward side of obstacle
c Tail dune on leeward side
d Wind direction

2 Barkhans
a Gentle windward face
b Steep leeward face
c Horns
d Prevailing wind
e Eddy

3 Seif dunes
A Seif dune pattern
a Seif dunes: long, narrow, straight, and parallel
b Corridors
c Prevailing wind
d Eddies

B Barkhan into seif
a Barkhan and prevailing wind
b Wind shifts: one horn lengthens
c Wind vacillates
d Later: seif takes shape

1 Head and tail dunes grow in dead air spaces near a rock or shrub. The long leeward tail can grow to almost half a mile (750 m) long.

2 Barkhans are dunes with low, curved flanks like horns, blown forward faster than the middle. Some barkhans grow 100 feet (30 m) high. Each migrates as wind pushes sand across its crest.

3 Seif dunes ("**sword dunes**") form long, wavy ridges up to 700 feet (215 m) high; they are thrown up by vortices in a prevailing wind or where barkhans are elongated by a cross wind.

Windblown particles finer than sand settle far beyond their point of origin. Loosely cemented silt-sized grains from Mongolia's Gobi Desert form layers up to 1,000 feet (300 m) thick in northern China. Geologists call these deposits loess, from a German word for "loose." Known as adobe in the USA and as *limon* in France, similar material covers parts of North America and Europe where winds blew dust from sands and clays deposited by ice sheets in the Pleistocene.

Loess deposits *(below)*
World map that locates loess and loess-like deposits in four continents and in New Zealand.
a North America
b South America
c Europe
d Asia
e New Zealand

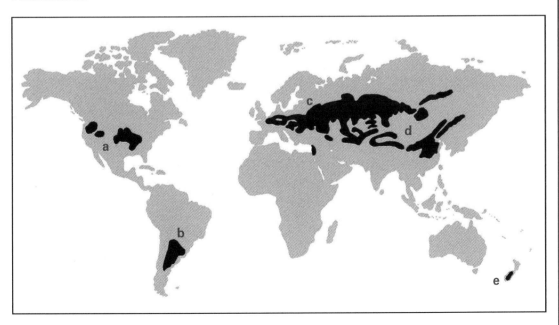

© DIAGRAM

9 | LANDS SHAPED BY WIND AND WATER

Desert weathering, flash floods, and/or windborne sand produce five main types of desert landscape: (**1**) jagged rock peaks, as in Sinai and the Sahara's Tibesti Mountains; (**2**) desert plateau with steep cliffs and deep, narrow river valleys; (**3**) stony, gravelly desert, also called *reg* or *serir*; (**4**) bare rock desert called a *hamada*; and (**5**) sand desert, also called *erg* or *koum*. Most have harsher features than you find in lands where vegetation softens the effects of sun, wind, frost, and rain. These five desert features figure mainly in such regions as the Colorado plateau:

1 Mesa Flat-topped, steep-sided plateau of horizontal strata capped by erosion-resistant rock.

2 Butte Isolated flat-topped hill, like a mesa but smaller.

3 Inselberg Steep, isolated hill with a narrow summit.

4 Pediment Gentle slope often covered with loose rock and lying below a mesa, butte, inselberg, or ridge. Pediments seem to be formed by weathering and floodwater.

5 Canyon Deep gorge of a river, often one flowing through a desert but fed by water from outside.

Eight desert features
1 Mesa
2 Butte
3 Inselberg
4 Pediment
5 Canyon
6 Wadi
7 Alluvial fan
8 Playa

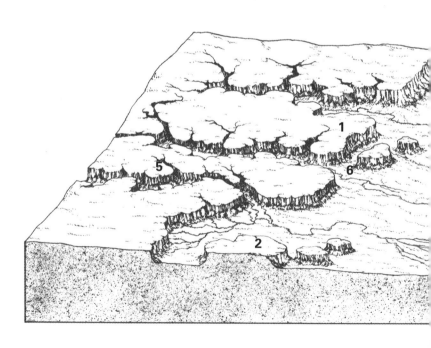

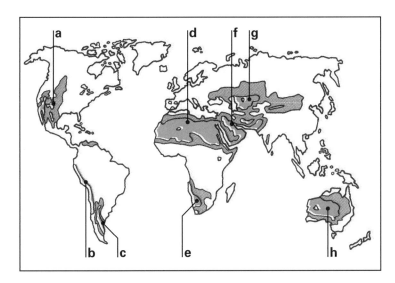

The world's deure

The world's deserts
Arid (**A**) and semiarid (**B**) areas lie mainly in the dry hearts of continents.
a North American desert
b Peruvian-Atacama desert
c Patagonian desert
d Sahara Desert
e Southwestern African deserts
f Middle Eastern deserts
g Central Asian deserts
h Australian desert

A
B

6 Wadi Usually dry desert watercourse, also called an *arroyo*, wash, or *nullah*.
7 Alluvial fan Fan-shaped mass of alluvial deposits shed by a fast-flowing mountain stream entering a plain or broad valley.
8 Playa (salt pan) Temporary brackish lake; many are found in Nevada and Utah.

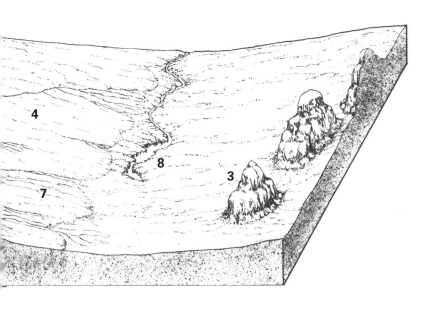

© DIAGRAM

CHAPTER 10 CHANGE THROUGH THE AGES

Processes producing and destroying land have shaped the crust ever since our planet developed its solid rocky skin. Geologist-detectives can now identify the broad sequence of events. Chronometric dating and relative dating based on layered rocks and fossils help scientists read the story in the rocks, but the first three of its four great volumes—time spans known as eons—are the least known and the longest.

The Grand Canyon, Arizona. (From *The United States of America* by E O Hoppé, 1928)

10 | RELATIVE DATING: USING ROCKS

A

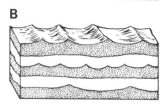

B

C

D

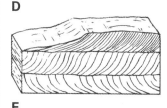

E

Clues to age sequences
Youngest rocks are at the top.
A Mud cracks
B Ripple marks
C Graded bedding
D Cross bedding
E Pillow lava

Earth's history lies locked up in the rocks that form its crust. Sedimentary rock layers, or strata, were laid down on top of one another, like pages in a history book. Reading these pages is the study called stratigraphy. Much of this is based on analyzing the properties of rocks themselves.

Geologists identify individual strata largely by such properties as grain size, minerals, and color. These features and distinctive fossils help experts to define rock units, or formations. Geologists may divide formations into members and beds or lump them together in subgroups, groups, or supergroups.

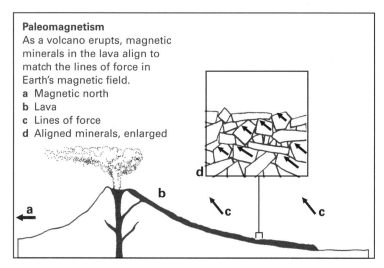

Paleomagnetism
As a volcano erupts, magnetic minerals in the lava align to match the lines of force in Earth's magnetic field.
a Magnetic north
b Lava
c Lines of force
d Aligned minerals, enlarged

Many pages in Earth's "book" are torn, turned upside down, displaced, or lost. Stratigraphy involves working out the true sequence in which rock strata formed in any given place, matching these with layers elsewhere, and noting local gaps where erosion has wiped strata from the record.

Various clues help rock detectives discover where Earth movements or injected molten rock have tampered with the evidence. For instance, steeply sloping strata never formed that way. Faults, folds, and injections of molten rock are younger than the rocks they affect. Mud cracks, ripple marks, and pillow lava create distinctive

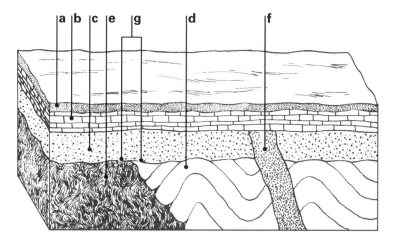

Which formed when? *(left)*
Volcanic ash (**a**) fell on limestone (**b**) laid down on glacial debris (**c**) dumped on slate (**d**) changed from preexisting shale by granite (**e**). A diabase (dolerite) dike (**f**) pierced **d** and **c** before **a** and **b** formed. Glacial erosion explains the unconformity at **g**.

patterns on a layer's upper surface, which becomes its base if the rock is overturned. A break between level rock layers above and crumpled layers below is an unconformity, suggesting a time gap when rock layers vanished by erosion.

Other clues help experts correlate the age of rocks formed at the same time in different places. Widely separated strata may share a unique set of characters, or facies. Migrating shorelines may mark a worldwide fall or rise in sea level. Widespread layers of volcanic ash could hint at an immense volcanic eruption. Lavas or sediments accumulating at the same time lock in particles aligned in the same direction by Earth's magnetic field which has undergone a sequence of reversals. Matched alignments and matched fossils help geologists to correlate the relative ages of rock cores sampled from around the world.

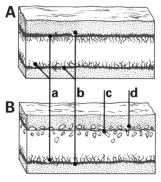

Sill or buried lava flow?
(above)
The following clues help geologists tell a sill (**A**) from a buried lava flow (**B**).
a Chilled margin
b Baked contact
c Weathered lava surface
d Eroded bits of lava in an overlying bed

Migrating shoreline *(left)*
Facies change vertically where rising sea level moved a shoreline to the right.
A Sea
B Beach
C Lagoon
a Limestone
b Fine sand
c Coarse sand
d Windblown sand
e Mud

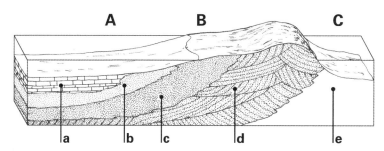

10 RELATIVE DATING: FOSSILS 1

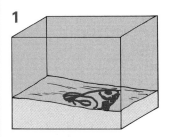

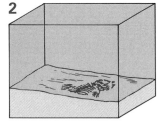

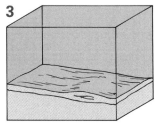

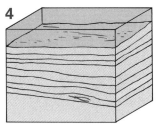

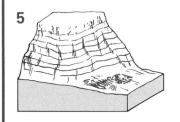

What we know of the relative ages of rocks owes much to paleontology—the study of fossils. Fossils are remains of prehistoric organisms, locked in sedimentary rocks being laid down when those organisms died.

Most dead organisms soon rot away. Fossils tend to be hard parts like wood or bone quickly buried and preserved by sediment below a sea or lake. Percolating minerals may reinforce a fossil. Or a corpse may dissolve to leave a hollow called a mold. Minerals that fill a mold create a cast. Plant leaves leave carbon films. Even burrows tracks, and droppings can be fossilized.

In time Earth movements lift fossil beds above the sea. Rivers cut down through the topmost bed, exposing lower layers. So paleontologists find fossils formed in rocks laid down at different times.

These studies show that living things evolved through time. Minute one-celled organisms gave rise to many-celled plants and animals. Some soft-bodied sea creatures led on to animals with shells or inner skeletons. Fishes with bony skeletons gave rise to amphibians, reptiles, birds, and mammals.

Through many generations tiny but accumulating changes in inherited characters altered organisms, better fitting them to feed, breed, and survive their enemies. So arose millions of species - twigs on a mighty tree of life. Related dead or living species form a genus. Related genera comprise a family. Related families belong to successively more major groupings— classes, phyla (animals) or divisions (plants), and kingdoms.

A fossil forming (left)
1 A dead fish on the sea bed.
2 Flesh rots, baring bones.
3 Mud or sand buries and preserves the bones.
4 Minerals harden bones now hidden by layered sediments.
5 Weather removes upraised rocks, exposing the fossil.

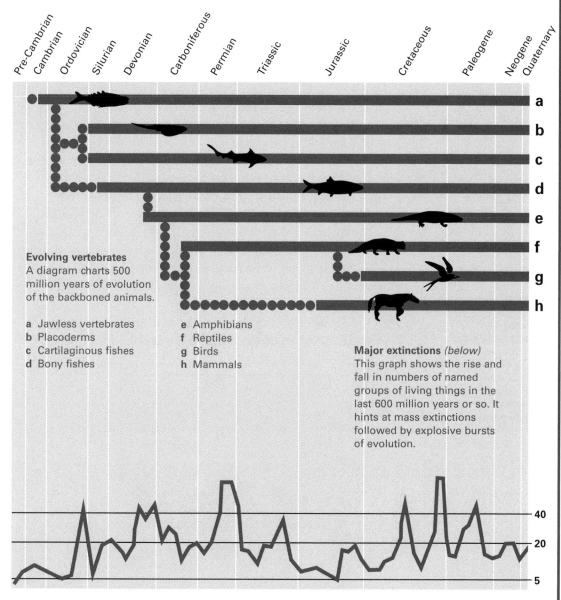

Pre-Cambrian · Cambrian · Ordovician · Silurian · Devonian · Carboniferous · Permian · Triassic · Jurassic · Cretaceous · Paleogene · Neogene · Quaternary

a
b
c
d
e
f
g
h

Evolving vertebrates
A diagram charts 500
million years of evolution
of the backboned animals.

a Jawless vertebrates
b Placoderms
c Cartilaginous fishes
d Bony fishes

e Amphibians
f Reptiles
g Birds
h Mammals

Major extinctions *(below)*
This graph shows the rise and
fall in numbers of named
groups of living things in the
last 600 million years or so. It
hints at mass extinctions
followed by explosive bursts
of evolution.

40
20
5

Most major groups are very old indeed. But there
survives a mere fraction of the hundreds of millions of
species evolving in the last 600 million years. New
enemies, or harsh environmental change destroyed the
rest. Fossil records in the rocks show waves of mass
extinction, then explosive evolution as new organisms
moved into habitats left empty by immense catastrophes.

RELATIVE DATING: FOSSILS 2

A

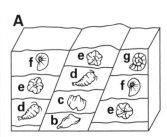

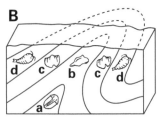

Fossils offer valuable aids to the relative dating of sedimentary rock strata and correlating these around the world. This process, biostratigraphy, involves identifying faunal zones—rock strata containing unique assemblages of fossils. Geologists name each faunal zone after a distinctive species called a zone fossil.

Besides providing guides to evolution and the ages of rocks, fossil individuals and groups reveal how prehistoric living things behaved and the kinds of place and climate they inhabited.

There are limits to our knowledge. Most soft-bodied organisms left no fossil record. Relatively few land plants and animals were fossilized. Billions of fossils vanished when erosion wore away the rocks containing them, or these were baked or crushed by metamorphic change. Billions more are inaccessible. But new kinds of fossil are discovered every year.

A good zone fossil meets four requirements: Its species was extremely plentiful; spread far and fast

Fossils, faults, and folds
(above)
Key fossils help geologists date rock layers disturbed by (**A**) faults or (**B**) folding.
a Cambrian trilobite
b Ordovician crinoid (sea lily)
c Silurian brachiopod (lamp shell)
d Devonian eurypterid (sea scorpion)
e Carboniferous blastoid (kin to starfish and sea urchin)
f Permian ceratite ammonoid
g Triassic ammonite

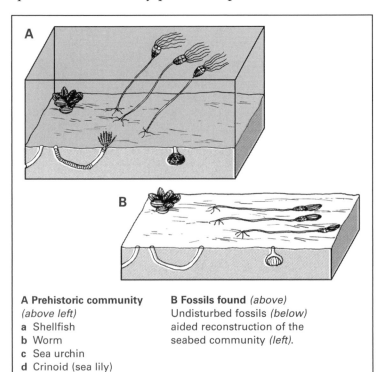

A Prehistoric community
(above left)
a Shellfish
b Worm
c Sea urchin
d Crinoid (sea lily)

B Fossils found *(above)*
Undisturbed fossils *(below)*
aided reconstruction of the
seabed community *(left)*.

(planktonic organisms are examples); left readily preserved remains; yet soon died out, thereby limiting its fossils to a few rock layers. Most such organisms lived in the sea. They ranged from sizable (macrofossil) animals and plants to tiny (microfossil) forms. Here are four examples:

1 Trilobites (three-lobed) were marine, segmented, distant relatives of wood lice; zone macrofossils for rocks 540–490 million years old.

2 Ammonoids were cephalopod mollusks with coiled, flat, wrinkled shells; zone macrofossils for rocks 370–65 million years old.

3 Bivalves are headless mollusks with hinged, two-part shells; zone macrofossils for rocks 370–65 million years old.

4 Foraminiferans are tiny one-celled protozoan organisms drifting in the seas and forming limy shells pierced by tiny holes; zone microfossils for rocks up to 65 million years old.

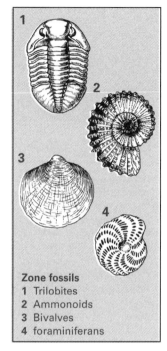

Zone fossils
1 Trilobites
2 Ammonoids
3 Bivalves
4 foraminiferans

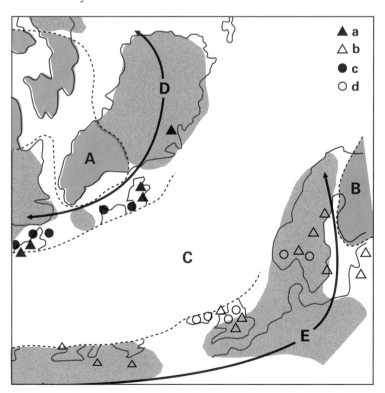

▲ a
△ b
● c
○ d

Faunal provinces *(left)*
Two groups of fossil trilobite (**a, b**) and graptolite (**c, d**) mark two faunal provinces–shelf seas flanking a pre–Atlantic Ocean before 500 million years ago.

A Land (proto-Greenland)
B Land (proto-Europe)
C Iapetus (pre-Atlantic) Ocean
D Pacific province
E Atlantic province

© DIAGRAM

177

CLOCKS IN ROCKS

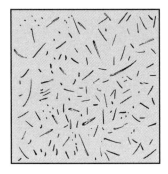

Fission-track dating
Fission tracks in a crystal are etched to make them show up, then counted under a microscope. The more tracks, the older the crystal.

Chronometric dating gives approximate ages in years for the rocks. Some rare sediments are datable from annually added layers. Some rocks are dated by the known rate of decay of a radioactive element into a more stable element. The more time that elapses, the less of the parent element remains and more of the daughter element accumulates. So measuring the proportions of both elements within a rock reveals its age. For igneous rock this means how long ago its minerals crystallized; for sedimentary rock, it indicates when sedimentation produced certain minerals; for metamorphic rock, it shows when heat drove daughter elements from the rock and "reset" the geological clock.

Dating diabase
Five illustrations depict six stages in dating a diabase (dolerite) dike injected into older rocks:

1 This cross section shows a diabase dike (a) injected into preexisting granite (b) before overlying sandstone (c) formed.
2 A lump of diabase is dropped into a crusher that grinds the rock until it breaks up into component mineral grains.
3 Froth flotation separates micas from other minerals.
4 A mass spectrometer using a magnetic field separates and measures the amounts of isotopes of potassium and argon.
a Gas inlet
b Electron beam
c Slits
d Ion accelerating voltage
e Ion beam
f Isotopes
g Detector slit
h Magnetic field

5 A computer printout of the isotope data enables the calculation of the diabase's potassium argon age. (Potassium is the parent element, and argon is the daughter element.)

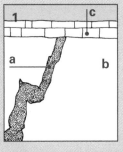

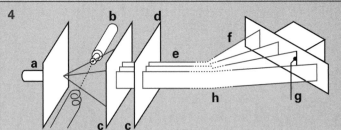

Various elements and their isotopes (different forms of atom of a given element) give best results in different circumstances. Here are brief details of three radiometric techniques and one based on nuclear fission:

1 Potassium-argon dating uses the decay of the potassium-40 isotope into argon-40. Applications: mainly igneous and metamorphic rocks of any age greater than about one million years, and sedimentary rocks containing the mineral glauconite.

2 Rubidium-strontium dating uses the decay of rubidium-87 to strontium-87. Applications: igneous and metamorphic rocks (except basic types), and sedimentary rocks containing the mineral illite. This method is often best for rocks more than 30 million years old.

3 Uranium-thorium-lead methods employ radioactive isotopes in uranium. Uranium-235 decays to lead-207; thorium-232 decays to lead-208. Applications: igneous intrusions, metamorphic rocks, and sediments containing zircon. These methods are best for rocks more than 100 million years old.

4 Fission-track dating requires counting the fission tracks produced in rock by the splitting nuclei of uranium-238, whose nuclei split at a known and constant rate. The older the rock, the more fission tracks there are. Applications: many igneous and metamorphic rocks.

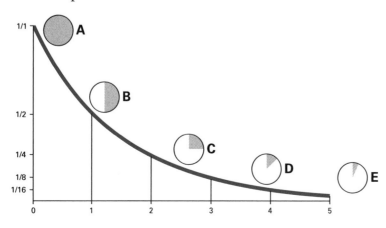

Radiometric dating *(left)*
By radiation potassium-40 loses half its mass every 1,310 million years (one half-life). Thus a sample's potassium-40 content can indicate its age.
A Original sample
B After 1.3 billion years (one half-life) half remains.
C After 2.6 billion years (two half-lives) one quarter remains.
D After 3.9 billion years (three half-lives) one eighth remains.
E After 5.2 billion years (four half-lives) one sixteenth remains.

© DIAGRAM

10 | THE GEOLOGICAL COLUMN

Grand Canyon *(left)*
From the rim of Arizona's Grand Canyon, spectators gaze down 6,250 feet (1,900 m) through more geological history than any other place on Earth. Exposed rocks date from two billion down to 220 million years ago.

By studying strata, geologists built up the geological column—a full record of Earth's crust's rocks laid down in sequence. This column is based on rock units, each comprising rock layers created in one geological period of time. In ascending order several of the time rock units, called zones, make up one stage; several stages form a series; several series build a system; and several systems make an erathem.

At first, geologists assigned only relative dates to geological time intervals. Radiometric dating occasionally enables them to be much more exact, yet dates are less precise the farther back we go.

Four eons *(right)*
A Hadean eon (4.5–4 billion years ago)
B Archean eon (4–2.5 billion years ago)
C Proterozoic eon (2.5 billion–543 million years ago)
D Phanerozoic eon (543 million years ago to Present)

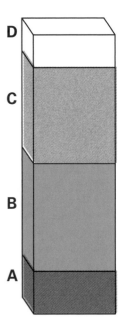

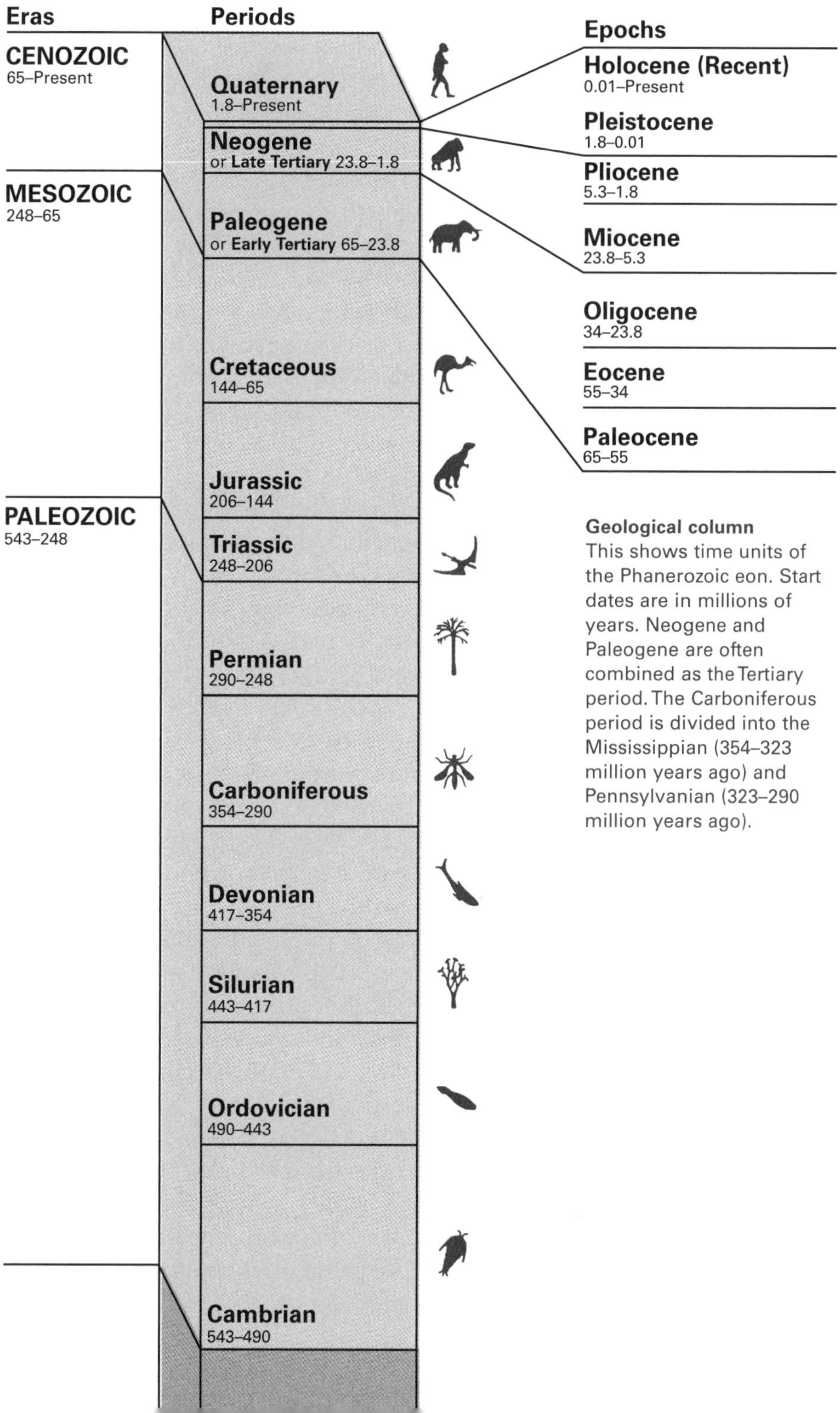

Eras	Periods	Epochs
CENOZOIC 65–Present	**Quaternary** 1.8–Present	**Holocene (Recent)** 0.01–Present
	Neogene or **Late Tertiary** 23.8–1.8	**Pleistocene** 1.8–0.01
		Pliocene 5.3–1.8
MESOZOIC 248–65	**Paleogene** or **Early Tertiary** 65–23.8	**Miocene** 23.8–5.3
		Oligocene 34–23.8
	Cretaceous 144–65	**Eocene** 55–34
		Paleocene 65–55
	Jurassic 206–144	
PALEOZOIC 543–248	**Triassic** 248–206	
	Permian 290–248	
	Carboniferous 354–290	
	Devonian 417–354	
	Silurian 443–417	
	Ordovician 490–443	
	Cambrian 543–490	

Geological column
This shows time units of the Phanerozoic eon. Start dates are in millions of years. Neogene and Paleogene are often combined as the Tertiary period. The Carboniferous period is divided into the Mississippian (354–323 million years ago) and Pennsylvanian (323–290 million years ago).

© DIAGRAM

181

10 | EARLY TIMES

The first continents (below) Archean minicontinents became the cores of modern continents, here shown as grouped about 250 million years ago.

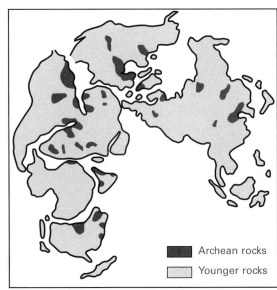

Archean rocks

Younger rocks

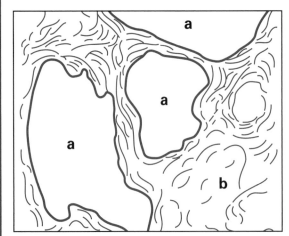

Archean rocks (above)
A satellite view shows these features.
a Cratons (granitoid continental nuclei)
b Greenstone belts

The first three eons of Earth's history are the longest and least known. The Hadean eon lasted from 4.5 to 4 billion years ago. Seemingly no rocks survive from the Hadean eon when the early Earth's crust was still molten. The Archean eon (Ancient Age) lasted from 4 to 2.5 billion years ago. The earliest-known surviving rocks on Earth are 3.9 billion years old. But they are older rocks reworked. The world probably already had some continental rock, an ocean, and an atmosphere—all produced by the resorting of Earth's less dense ingredients. Geologists do not agree how that occurred—perhaps as follows:

The crust and mantle were probably more active at that time than today. Where two cooling mantle currents met and sank, they squashed, thickened, and melted the thin primeval crust above. Repeated melting could have resorted its ingredients until the lightest formed a scum of continental igneous rocks that metamorphosed with others into gneisses. There maybe appeared "granitoid" microcontinents in this way, with some still surviving as the ancient cores of continents.

Wrapped around these microcontinents were greenstone belts of lightly metamorphosed greenish dark volcanic rock, combined with shales and sandstones. The volcanic rock came perhaps from volcanoes spewing lava, ash, and gas from scores of hot spots in a weak, thin, early crust. Or they were island arc volcanoes, and the shales and sandstones formed from sediments washed off nearby microcontinents. Perhaps volcanic steam that cooled and turned to rain filled

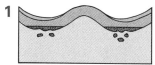

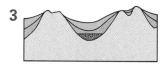

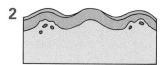

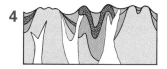

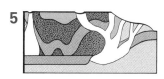

Continental growth
A Increasing sediment volume in the last 4 billion years implies slow growth of land.
B Continental sediments (**a**) and volcanic rocks (**b**) increased in volume, while greenstone belt volcanic rocks (**c**) and sediments (**d**) decreased. The scale shows millions of years ago.

early ocean basins and the rivers that eroded rocks. Certainly volcanic gases formed an early atmosphere, rich in nitrogen and carbon dioxide.

Fossil organisms in the greenstone belts reveal that life appeared at least 3.5 billion years ago. Bacteria and blue-green algae were flourishing in shallow seas. Oxygen released each year by algae combined with iron, producing the banded iron formations now mined around the world. But iron and other chemical sponges left little oxygen for adding to the atmosphere.

Stromatolites (below)
Blue-green algae formed intertidal pedestals like these, 3.5 billion years ago.

Possible processes (above)
These stages might have formed Archean microcontinents.
1 Sediments from uplifted crust fills sags in oceanic crust.
2 Melting forms acid magma.
3 Subsidence and melting form alkaline magmas.
4 Folding yields granitic magmas and greenstone belts.
5 The cooled product is a sialic shield and mobile belt.

© DIAGRAM

THE AGE OF FORMER LIFE

About 2.5 billion years ago the first large continents appeared, and extensive, shallow, offshore seas gave new opportunities for living things to develop. These changes ushered in the third phase of Earth's history. The Proterozoic eon (Age of Former Life) spanned 1.9 billion years, ending about 540 million years ago New "granitoid" and greenstone belts emerged, and vast masses of volcanic rocks and sediments were tacked on to Archean microcontinents.

Geologists detect three main construction phases, starting 1.9 billion, 1.2 billion, and 700 million years ago. The first produced the Wopmay Orogen—a long-since beveled mountain belt in northern Canada. Such ancient sites hold traces of old oceanic crust, island arcs, and colliding continents. This shows that plate tectonics was already molding continental and oceanic crust. Alignments of magnetic particles in rocks prove that continents were drifting and ocean floor was rifting and subducting by 1.5 billion years ago. Some evidence suggests that major continents were even stuck together at that time.

Meanwhile, the oceans and the atmosphere were undergoing change. Salts washed off land gave the sea its present saltiness. By about two billion years ago, cyanobacteria produced enough free oxygen for some to start accumulating in the sea and atmosphere. Proof of this comes from compounds formed in certain rocks.

The Wopmay Orogen (below) Colliding Bear plate (**A**) and Slave plate (**B**) crumpled intervening rock zones (**1–4**). Right: A map shows the same zones and the old Slave Plate all now in Canada south of the Arctic Ocean (**C**).

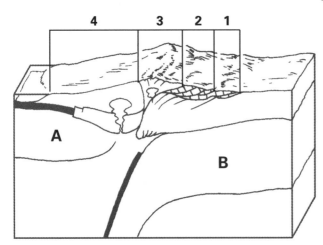

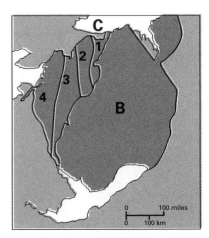

Atmospheric oxygen began to build an ozone shield protecting living things from the Sun's lethal ultraviolet radiation. New, complex kinds of water life evolved, able to exploit the energy in oxygen. Soon after 700 million years ago, soft corals, jellyfish, worms, and other soft-bodied animals perhaps flourished in shallow seas off continental shores.

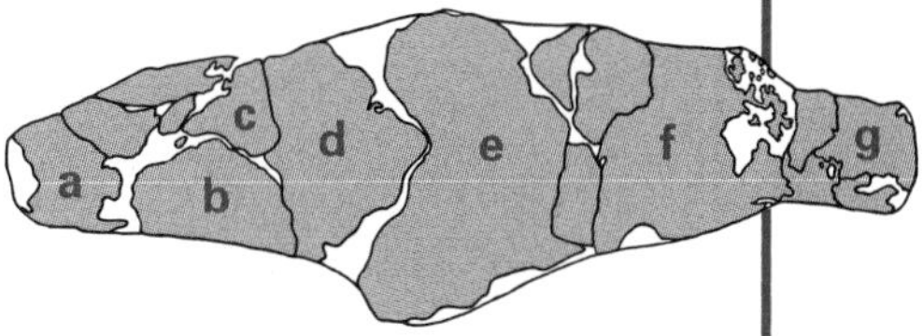

A Supercontinent *(above)*
A supposed mid-Proterozoic supercontinent would have held bits of future Asia and these recognizable land masses.
a Australia
b Antarctica
c India
d South America
e Africa
f Proto-North America
g Proto-Europe

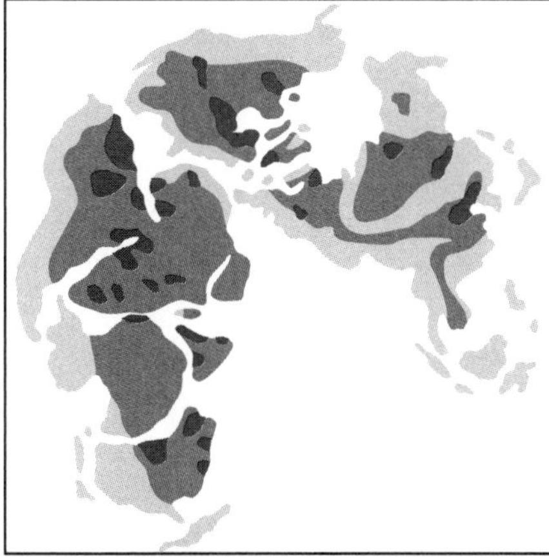

Expanding continents
Proterozoic rocks added new land to the old, Archean minicontinents. Lands appear as grouped 250 million years ago.

Younger

Proterozoic

Archean

Complex life
Late Proterozoic seafloor organisms of Australia:
a Ediacaria
b Mawsonites
c Pteridinium
d Charniodiscus
e Tribrachidium
f Parvancorina
g Dickinsonia
h Spriggina
i Algae

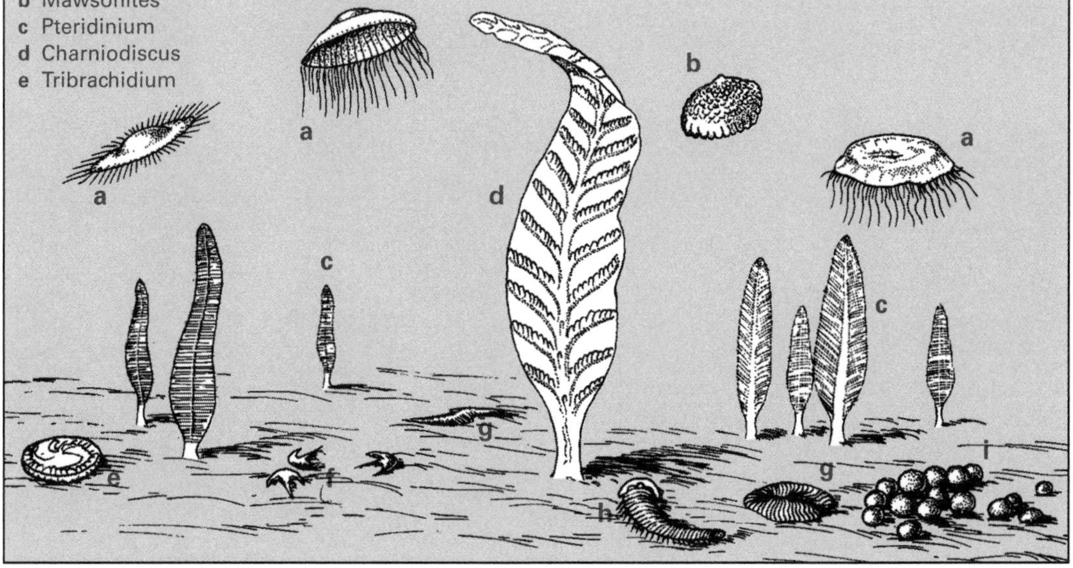

© DIAGRAM

185

CHAPTER 11 THE LAST 543 MILLION YEARS

These pages trace major events of the Phanerozoic eon—the last, shortest, and best recorded volume in Earth history. Era by era, period by period, we examine the vast changes that transformed our planet's continents, climates, and living things.

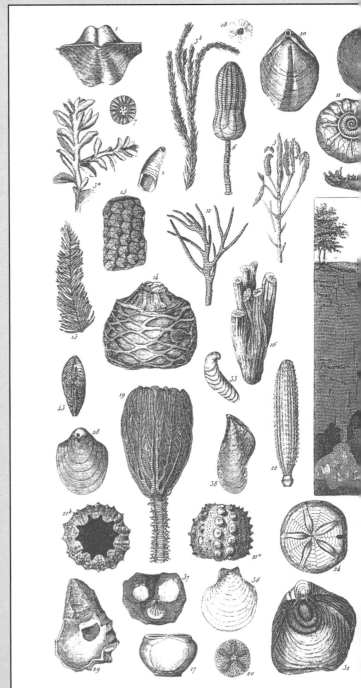

A section of the Wirksworth Cave, Derbyshire, England, and some of the fossils found there. (Engraving originally published in *The Iconographic Encylopaedia of Science, Literature and Art,* 1851)

11 | THE AGE OF VISIBLE LIFE

The last 543 million years or so form the Phanerozoic eon, or Age of Visible Life. The Phanerozoic saw complex modern life forms in the making.

Assemblages that lived at different times have led geologists to split the eon into three successive eras: the Paleozoic, or Age of Ancient Life (about 543–248 million years ago); Mesozoic, or Age of Middle Life (about 248–65 million years ago); and Cenozoic, or Age of Recent Life (about 65–0 million years ago). Fossil assemblages differing in time and place help experts reconstruct each era's climates, lands, seas, and oceans.

By coordinating fossil clues with those left by the rocks themselves, geologists have reconstructed how and why continents drifted, seas and oceans spread and shrank, mountain ranges rose and were worn down, and ice sheets waxed and waned.

Much remains unknown. For instance paleomagnetism reveals the past north-south positions of the continents, but their longitudes (east-west locations) lie open to dispute.

Even so, the following pages stress key events in each era's successive periods—the later chapters in the story of Earth.

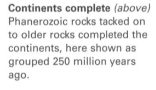

Phanerozoic
Proterozoic
Archean

Continents complete *(above)* Phanerozoic rocks tacked on to older rocks completed the continents, here shown as grouped 250 million years ago.

Continents drift *(below)* Segments show landmasses split, fused, and split again in the last 600 million years.

1 Early Paleozoic
2 Later Paleozoic
3 Latest Paleozoic
4 Mid Mesozoic
5 Cenozoic
a North America

b Europe
c Asia
d Gondwana
e Pangaea
f Tethys Sea
g South America

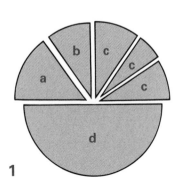

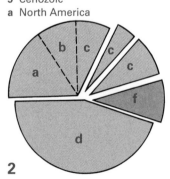

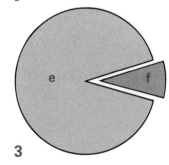

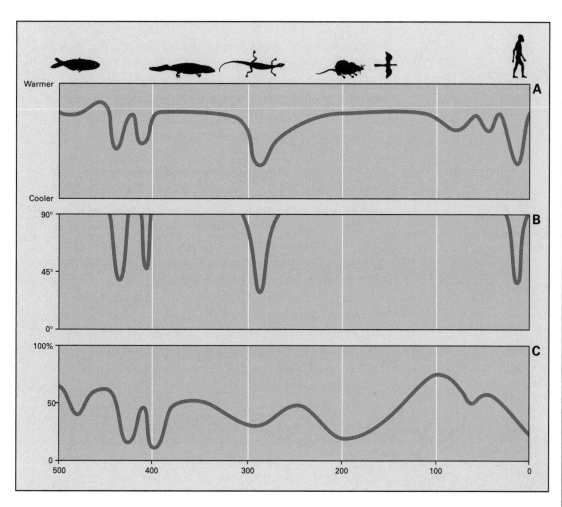

h Eurasia
i Antarctica/Australia
j Africa
k India
l Antarctica
m Australia

Climatic changes
Three graphs show climatic
and related changes in the
last 500 million years.
A Likely rise and fall in world
temperatures
B Glaciers' advance toward
and retreat from the
equator, at latitude 0°
C Spread and shrinkage of
continental seas

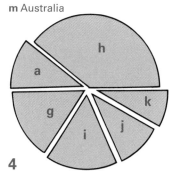

4

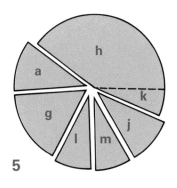

5

© DIAGRAM

11 CAMBRIAN PERIOD

Cambrian world *(above)*
Lands might have been arranged like this. Lines show the equator, tropics, and polar regions.

Cambrian rocks exposed *(below)*
Cambrian limestone overlying Cambrian quartzite form these high cliffs in Canada's Banff National Park. The cliffs' materials were laid down as immensely thick deposits in shallow water off what was then the northwest rim of proto-North America.

The Cambrian period (about 543–490 million years ago) takes its name from the Latin word for Wales. Here geologists first studied Cambrian fossils—the first abundant animals with skeletons and shells. All life inhabited water. It teemed in shallow seas that invaded continents as ice sheets melted about 600 million years ago. Cambrian times were generally warmer than today.

Most continental lands probably lay on or close to the equator. South America, Africa, India, Antarctica, Australia, and bits of Asia were evidently welded into one southern supercontinent—Gondwana—with Africa "upside down." Among lesser chunks of continental crust were the cores of North America, Greenland, Europe, and northwest Africa. The Iapetus Ocean—a pre-Atlantic Ocean—had opened up between these once-fused lands. Within this ocean lay Avalonia, an archipelago whose rocks today lie scattered from the Carolinas, north through Newfoundland, to parts of Ireland and Wales.

By 570 million years ago mountains had sprouted in the Avalonian orogeny when a slab of land struck

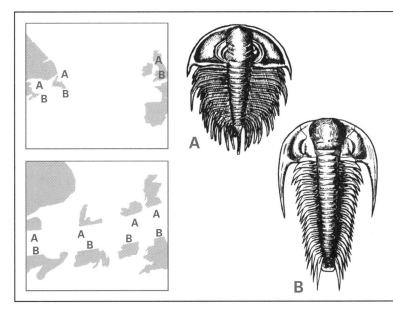

A trilobite tale
Two Cambrian trilobites (right) posed a geologic puzzle. Fossil Olenellus (**A**) occurs as seen in Map 1— in northern parts of Nova Scotia, Newfoundland, and the British Isles (shown closer than they really are). Fossil Paradoxides (**B**) occurs farther south in all three places. Map 2 shows why this split exists. In Cambrian times each place was two areas of shallow sea floor separated by the Iapetus Ocean. This deep-sea barrier kept both kinds of trilobite apart.

eastern North America to form New England. Elsewhere, the Andes mountains had begun to grow, and volcanoes spewed vast sheets of lava over parts of north and west Australia.

Sands that washed off land into shallow seas provided raw materials for sandstones. Animals with shells became a source of carbonates, and dolomite began to form. Other Cambrian sediments include blue clays that still survive in Russia.

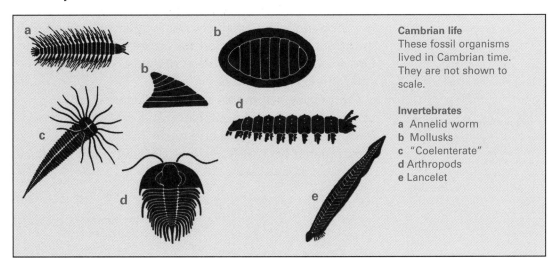

Cambrian life
These fossil organisms lived in Cambrian time. They are not shown to scale.

Invertebrates
a Annelid worm
b Mollusks
c "Coelenterate"
d Arthropods
e Lancelet

© DIAGRAM

ORDOVICIAN PERIOD

Ordovician world *(above)*
Lands might have been arranged like this. Lines show the equator, tropics, and polar regions.

Rocks from this time (about 490–443 million years ago) were first studied in Wales; an early Welsh tribe, the Ordovices, inspired the period's name.

Subduction brought slow shrinkage of the Iapetus Ocean and a closing up of its flanking continental cores: Laurentia (proto-North America and Greenland), Baltica (proto-Europe), and northwest Africa. Early on, subducting or colliding crustal blocks deformed rocks of the future Scottish Highlands. Later, the Taconic orogeny forced up the Green Mountains of Vermont. Volcanic rocks appeared in Scandinavia and Greenland.

Although large slabs of continental crust lay close to the equator, part of Gondwana moved deep into Antarctic latitudes. Indeed northwest Africa lay astride the South Pole. Late Ordovician ice-scoured rocks and ice-borne debris show that ice sheets covered northwest Africa and nearby parts of South America.

At their maximum extent the ice sheets locked up much of the world's water, and shallow continental seas withdrew. But from time to time, warm, shallow, salty water invaded lowlands including parts of proto-North America. Here lived trilobites, early corals, and many more invertebrates. The small colonial organisms called graptolites left fossils used for correlating the ages of Ordovician rocks from different areas. There were early jawless fishes, too.

Ordovician sediments up to 23,000 feet (7,000 m) thick formed in continental seas and offshore waters. North America, Europe, and north Australia all accumulated beds of limestone, dolomite, and coral.

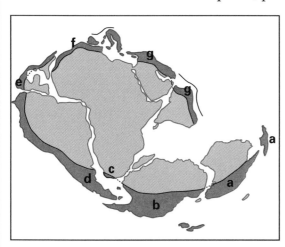

Future mountains *(above)*
Rocks eroded from Gondwana rimmed that land with belts of sediments in downwarped crustal troughs—later a source of mountain chains and ranges.
a Tasman Mio-/Eugeocline
b Transantarctic Geocline
c Cape Geocline
d Andean Mio-/Eugeocline
e Appalachian Geocline
f West African Geocline
g Southern Tethyan Geocline

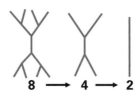

$$8 \longrightarrow 4 \longrightarrow 2$$

Evolving graptolites *(left)*
Ordovician times saw a reduction in the number of branches formed by colonies of tiny sea creatures, called graptolites.

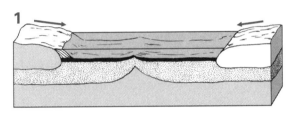

Ordovician life
These fossil organisms lived in Ordovician time. They are not shown to scale.
Invertebrates
a "Coelenterate"
b Mollusks
c Arthropod
d Brachiopod
e Echinoderms
f Graptolite

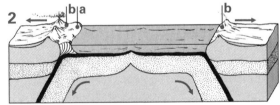

Taconic orogeny *(left)*
1 Cross section through the Iapetus Ocean in Cambrian times
2 Subduction of the Iapetus crust might have raised sediments (**a**) and spawned volcanoes (**b**), forming mountains on both shores.

Ordovician rocks exposed
(lower left)
The shales and sandstones above this road at South Africa's Cape of Good Hope contain late Ordovician fossils and pebbles scratched by Ordovician glaciers.

© DIAGRAM

11 | SILURIAN PERIOD

Silurian world *(above)*
Lands might have been arranged like this. Lines show the equator, tropics, and polar regions.

1

2

3

Silurian life *(above)*
1 Cystiphyllum, a solitary coral
2 Baragwanathia, a lycopsid—an early land plant
3 Palaeophonus, a scorpion—one of the first land animals

The Lower Paleozoic ended with the Silurian period (443–417 million years ago). The Silures were an ancient British tribe of the Welsh border region where geologists first studied Silurian rocks. Rocks dating from this time occur on almost every continent. Some contain very early fossil land plants and animals.

As the pre-Atlantic Iapetus Ocean shrank, proto-North America and Greenland were beginning to collide with proto-Europe. The Caledonian orogeny crumpled up the edges of their plates, pushing up forerunners of the Scandinavian, Caledonian (Scottish), and Appalachian Mountains—a mighty chain of peaks that would extend in time from Scandinavia through the British Isles and Greenland to New York. Farther west, North America ended at eastern Nevada and Idaho. But as the continent began to override the oceanic plate beyond, new land would stick on to its western rim.

Some experts think Asia was mainly three ocean isolated blocks—Siberia, China, and part of South-East Asia. But the first two were closing by Silurian times. Indeed perhaps all northern continental slabs had almost fused to form a northern supercontinent, Laurasia.

In the southern supercontinent, Gondwana, Africa and South America were drifting north while Antarctica and Australia still headed south.

Silurian rocks exposed *(below)*
These Silurian grits on the west Welsh coast formed horizontal layers until tilted by the Caledonian orogeny.

Melting southern ice sheets flooded continents with shallow seas. Off proto-North America the seafloor gained thick sheets of sands and gravels, the eroded ruins of high mountains raised in Ordovician times. Other sediments produced rich oil reserves in what is now the Sahara Desert. Widespread reefs marked the spread of (solitary) corals, and evaporites accumulated on the arid western coasts of continents.

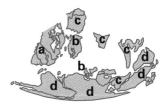

Alternative view (above)
Some experts think that today's northern continents remained largely unassembled.
a Bits of North America
b Bits of Europe
c Bits of Asia (perhaps even more than shown)
d Southern continents with bits of northern ones

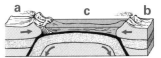

A closing ocean (above)
Proto-North America (**a**) and proto-Europe (**b**) advance on one another, closing the Iapetus Ocean (**c**) and starting the Caledonian orogeny.

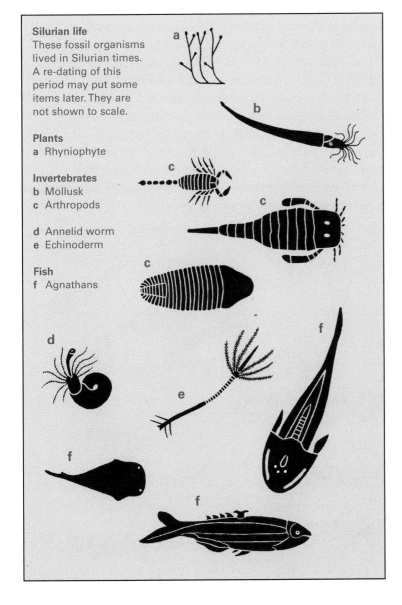

Silurian life
These fossil organisms lived in Silurian times. A re-dating of this period may put some items later. They are not shown to scale.

Plants
a Rhyniophyte

Invertebrates
b Mollusk
c Arthropods

d Annelid worm
e Echinoderm

Fish
f Agnathans

© DIAGRAM

DEVONIAN PERIOD

Devonian world *(above)*
Lands might have been arranged like this. Lines show the equator, tropics, and polar regions.

The Devonian period takes its name from Devon, England, where shales, slates, and Old Red Sandstone were laid down about 417–354 million years ago. But every continent has rocks dating from this first phase of the Upper Paleozoic. Devonian deposits include widespread coral reefs and rich oil reserves in Canada and Texas.

During the Devonian, subducting oceanic crust and colliding northern continents entirely closed the northern Iapetus Ocean. The Acadian orogeny uplifted much of northeast North America while the Caledonian orogeny was still affecting Europe. Eastern North America, Greenland, and western Europe fused to form an Old Red Continent. Its Old Red Sandstone rocks are formed from the eroded fragments of the huge mountain chain thrown up by the collision. Fossils in such rocks include freshwater fishes and the first "amphibians," whose Greenland home then straddled the equator. Devonian rocks also hold fossil remnants of the world's first forests.

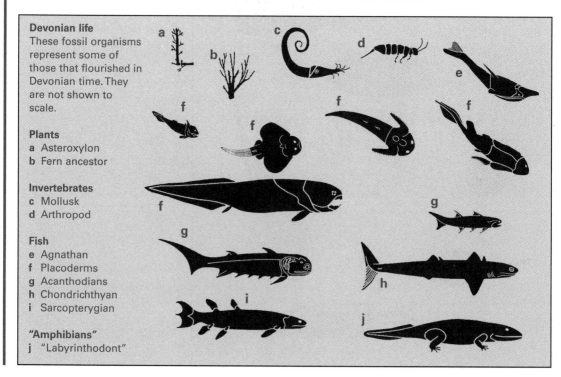

Devonian life
These fossil organisms represent some of those that flourished in Devonian time. They are not shown to scale.

Plants
a Asteroxylon
b Fern ancestor

Invertebrates
c Mollusk
d Arthropod

Fish
e Agnathan
f Placoderms
g Acanthodians
h Chondrichthyan
i Sarcopterygian

"Amphibians"
j "Labyrinthodont"

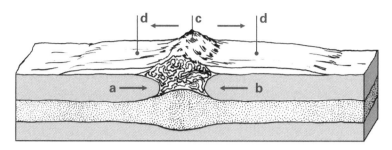

Caledonian-Acadian orogeny *(left)*
The colliding plates of proto-
North America (**a**) and proto-
Europe (**b**) closed the Iapetus
Ocean, crumpling intervening
rocks (**c**) into a chain of peaks
from Scandinavia to New
York. Their eroded remnants
formed Old Red Sandstone
(**d**).

At this time, Gondwana was moving north and pushing minicontinents ahead of it. Only a narrow, shrinking sea, the Tethys, separated South America and Africa from North America and Europe. Meanwhile the ocean separating Russia from Siberia was evidently closed. New collisions were inevitable. One theory holds that most of western Europe was created when the minicontinent Armorica slammed into Baltica, the proto-European continent. In Devonian times, this impact was foreshadowed by heavings that began pushing up the Hercynian mountain belt whose remnants include the Armorican Massif, Vosges, and Black Forest.

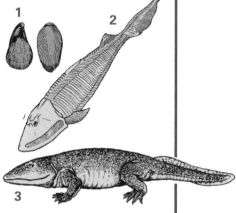

Devonian life *(above)*
1 Brachiopods (lamp shells)
2 Hemicyclaspis, a jawless fish
3 Ichthyostega, an early "amphibian," 3 feet 3 inches (1 m) long

Devonian rocks exposed *(below)*
Gently dipping Upper Old Red Sandstone beds overlie more steeply dipping Lower Old Red Sandstone in eastern Scotland.

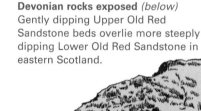

© DIAGRAM

11 CARBONIFEROUS PERIOD

Carboniferous world *(above)*
Lands might have reached these positions. Lines show the equator, tropics, and polar regions.

The Carboniferous period (354–290 million years ago) takes its name from thick, coal-producing carbon layers. These are the remains of swampy tropical forests that were submerged when shallow seas invaded a vast, low-lying tract embracing much of North America and Europe. Smaller forests flourished in South America and Asia.

In North America this period is split in two. The Mississippian (354–323 million years ago) saw limestones laid down by a shallow sea that covered the Mississippi region. The Pennsylvanian (323–290 million years ago) is named for coal measures formed in Pennsylvania about 323–290 million years ago. Coal forests then flourishing in Nova Scotia contained the first known reptiles.

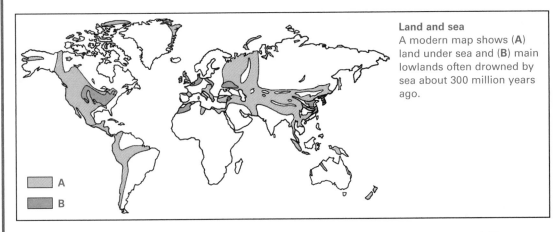

Land and sea
A modern map shows (**A**) land under sea and (**B**) main lowlands often drowned by sea about 300 million years ago.

A

B

The Carboniferous saw North America and Europe colliding with the northern edges of Gondwana—the part containing South America and Africa. By about 350 million years ago, this process was fusing northern and southern continents into a single landmass, called Pangaea.

The slow collisions forced up mountain ranges. About 300 million years ago South America seemingly struck Texas and Oklahoma, pushing up the Ouachita Mountains. Later, South America or Africa smashed into southeast North America. This resulted in the Alleghenian orogeny, crumpling up the southern

Appalachians. Meanwhile the collision of South America or Africa with the south of Europe destroyed the intervening sea and continued raising the Hercynian mountains, whose eroded roots still run from southern Ireland to Bohemia. Such impacts also generated mountains in Gondwana.

As drifting carried parts of southern continents across the South Pole, ice sheets again smothered regions of the Southern Hemisphere. By late Carboniferous times, ice covered all of Antarctica, parts of Australia, and much of southern South America, Africa, and India.

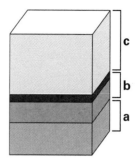

A sediment sequence *(above)* Non-marine sandstone (**a**), transitional sediments (**b**), including coal, and marine sediments (**c**) formed in recurring sequence in late Pennsylvanian times.

Carboniferous life *(left)*
1 Conodont animal, a tiny sea creature, enlarged. Conodonts, or "cone-teeth," supported soft tissue in the gut area.
2 Meganeura, a giant proto-dragonfly
3 Hylonomus, an early reptile, was about 3 feet 3 inches (1 m) long.

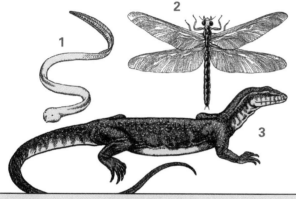

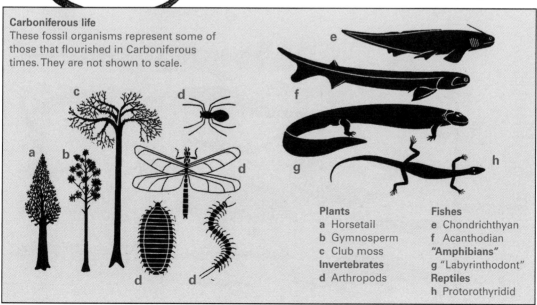

Carboniferous life
These fossil organisms represent some of those that flourished in Carboniferous times. They are not shown to scale.

Plants
a Horsetail
b Gymnosperm
c Club moss
Invertebrates
d Arthropods

Fishes
e Chondrichthyan
f Acanthodian
"Amphibians"
g "Labyrinthodont"
Reptiles
h Protorothyridid

© DIAGRAM

11 PERMIAN PERIOD

Permian world (above)
Lands might have looked like this, omitting (much shrunken) continental seas. Lines show the equator, tropics, and polar regions.

The Permian period (about 290–248 million years ago) is named after rocks from the old province of Perm in Russia's Ural Mountains.

This wall between northern Europe and Asia sprang up in Permian times when the Siberian plate collided with eastern Russia. Meanwhile other "pre-Asian" plates were quite likely docking and deforming the rocks of intervening mobile belts in forging most of the rest of Asia.

Elsewhere, Africa's (or South America's) collision with Europe's southern underbelly buckled up the mountains of Europe's Hercynian mobile belt. Farther west, Africa's (or South America's) collision with southeast North America went on crumpling up the southern Appalachians. On several continents collisions cracked open the crust, releasing basalt lavas.

Finally all continents lay jammed together as the supercontinent Pangaea, surrounded by the single, mighty Panthalassa Ocean.

Permian life
These fossil organisms represent some of those that flourished in Permian times. They are not shown to scale.

Plants
a Tree fern
b Conifer
Invertebrates
c Arthropods

Fishes
d Chondrichthyan
e Bony fish
"Amphibians"
f "Labyrinthodonts"
g Lepospondyls.

Reptiles
h Captorhinid
i Mesosaur
j Therapsid
k Pelycosaur
l Lepidosauromorph
m Pareiasaur

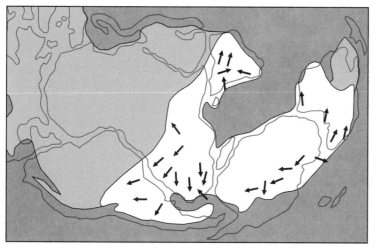

Permian ice sheets *(left)*
This map of early Permian Gondwana (fused southern continents) suggests that moving ice sheets covered much of South America, southern Africa, India, Antarctica, and southern Australia. Scoured rocks and glacial deposits hint at ice flow and its extent.

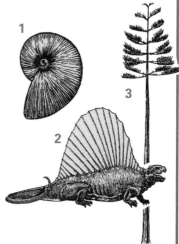

As Pangaea drifted north, glaciers retreated south in South America, Africa, and India. They gripped Antarctica as that landmass crossed the South Pole, as well as areas of Australia.

With much water locked up in ice and uplift of some continental masses, continental seas began to drain away. Large tracts of northern continents experienced a dry, continental climate with deserts that contained evaporites. Iron-rich minerals strongly oxidized in warm conditions produced the vivid rusty red beds typical of sedimentary rocks laid down in these conditions.

As the Permian period (and Paleozoic era) closed, the loss of many continental seas and vast volcanic eruptions helped produce the greatest mass extinctions in the fossil record.

Permian life *(above)*
1 Medlicottia, an ammonoid
2 Dimetrodon, a flesh-eating reptile 11 feet 6 inches (3.5 m) long
3 Conifer, a tree bearing seeds in cones

Permian mountains *(left)*
Russia's Ural Mountains were thrust up where eastern Europe docked with western Asia.

© DIAGRAM

TRIASSIC PERIOD

Triassic world (above)
Lands might have looked like this, omitting continental seas. Lines show the equator, tropics, and polar regions.

Triassic life (right)
1 Tropites, a ceratite ammonoid
2 Gerrothorax, an "amphibian" 3 feet 3 inches (1 m) long
3 Cynognathus, an advanced mammal-like therapsid 5 feet (1.5 m) long

The Mesozoic era, often called the Age of Dinosaurs, opened with the Triassic period, dating from about 248–206 million years ago. The name "Triassic" comes from the Latin *trias* ("three"), it is derived from three rock layers found in Germany.

Scarce marine sediments suggest a low sea level early in Triassic times, but red beds and evaporites accumulated on the land.

Between the mid-Permian and mid-Triassic periods, the Pangaean landmass drifted north about 30 degrees. North America, Europe, and northwest Africa seemingly lay locked together, but not, perhaps, immovably. Some experts think a 2,200-mile (3,500 km) east-west shearing of northern continents in relation to South America and Africa brought North America and Europe closer to the positions they occupy today. A gulf—the Tethys Sea—separated southern Eurasia from Afro-India.

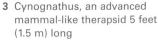

Cracked supercontinent (left)
The Palisades along New York's Hudson River are a Triassic or early Jurassic sill 400 feet (120 m) high. Its molten diabase (dolerite) rock rose through a crustal rift foreshadowing the break up of Pangaea.

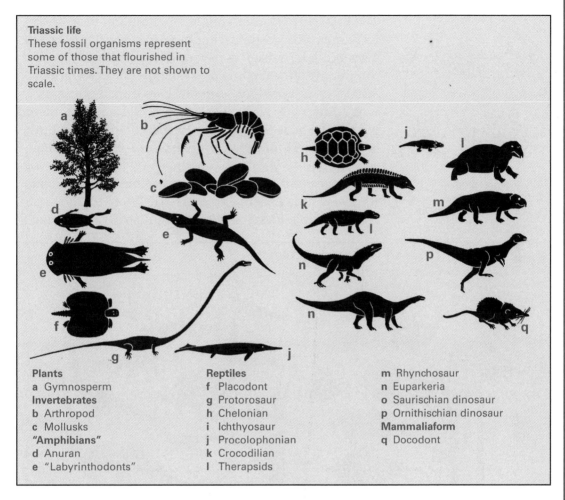

Triassic life
These fossil organisms represent some of those that flourished in Triassic times. They are not shown to scale.

Plants
a Gymnosperm
Invertebrates
b Arthropod
c Mollusks
"Amphibians"
d Anuran
e "Labyrinthodonts"

Reptiles
f Placodont
g Protorosaur
h Chelonian
i Ichthyosaur
j Procolophonian
k Crocodilian
l Therapsids

m Rhynchosaur
n Euparkeria
o Saurischian dinosaur
p Ornithischian dinosaur
Mammaliaform
q Docodont

Much of North America and Europe still lay inside the tropics. Gondwana no longer straddled the South Pole, and southern ice sheets had completely melted. World climates ranged from warm to mild, and deserts were extensive.

Pangaea now showed signs of breaking up. Here and there, rising plumes of matter in the mantle domed the crust above until it split, creating block faults leaking lava. Indeed, such rifts dated back to Carboniferous times in Scotland's Midland Valley and a Permian rift opened up in Norway. Triassic rifts affected west and central Europe, eastern North America, and northwest Africa. But Pangaea's true destruction lay ahead.

© DIAGRAM

JURASSIC PERIOD

Jurassic world *(above)*
Lands might have looked like this, omitting continental seas.

The Jurassic period (about 206–144 million years) takes its name from fossil-bearing limestone rocks formed in a sea but later raised as part of Europe's Jura Mountains. World climates were now mostly warm, and lands largely low, with old Paleozoic mountains worn down into stubs. Dinosaurs could have wandered overland across the world they shared with early birds and mammals. But the continental crust was growing restless.

Pangaea had started breaking up into the continents we know today. Here and there crust domed, then split, creating triple-junction rifts. Later, linked rifts formed a

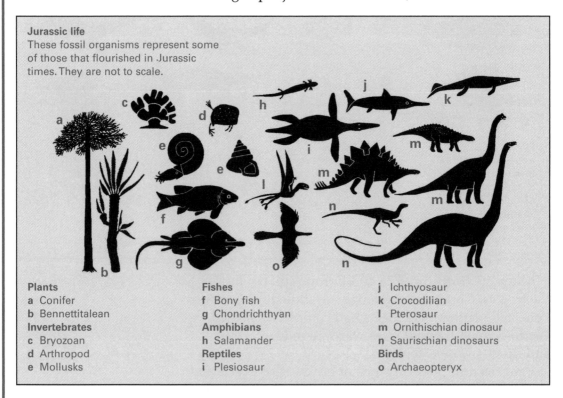

Jurassic life
These fossil organisms represent some of those that flourished in Jurassic times. They are not to scale.

Plants
a Conifer
b Bennettitalean
Invertebrates
c Bryozoan
d Arthropod
e Mollusks

Fishes
f Bony fish
g Chondrichthyan
Amphibians
h Salamander
Reptiles
i Plesiosaur

j Ichthyosaur
k Crocodilian
l Pterosaur
m Ornithischian dinosaur
n Saurischian dinosaurs
Birds
o Archaeopteryx

spreading ridge that opened up the central part of the Atlantic Ocean, divorcing eastern North America from northwest Africa. Bits of the original continents became transposed; North America probably gained Florida from Africa. (Elsewhere, Asia would eventually gain Siberia's eastern tip from North America.)

Meanwhile, rifting was separating Africa/South America from Antarctica/Australia, although the Indian subcontinent was probably still stuck to eastern Africa. As cracks appeared vast flows of molten basalt welled up from southern Africa, through Antarctic mountains to Tasmania. More basalt flows show Australia preparing to cast off from Antarctica.

About 145 million years ago, Africa pushed east against southern Europe, shedding crustal chunks. These minicontinents eventually formed parts of Spain, Italy, Greece, Turkey, Iran, and Arabia. Meanwhile Eurasia had been fusing with Tibet.

Moving west to override the ocean floor, western North America produced three mountain-building episodes. Volcanoes sprouted as far south as the central Andes. Mighty blobs of molten granite bobbed up, melting solid rocks above. And slabs of crust—some possibly from Asia—were jammed against the western rim of North America.

A limestone house *(above)*
Jurassic limestone provided walls for this fine old English Cotswold manor house.

Life on land *(above)*
1 Williamsonia, a plant with palmlike fronds
2 Diplodocus, a giant saurischian dinosaur with four limbs, long neck, and long tail
3 Archaeopteryx, the first known bird

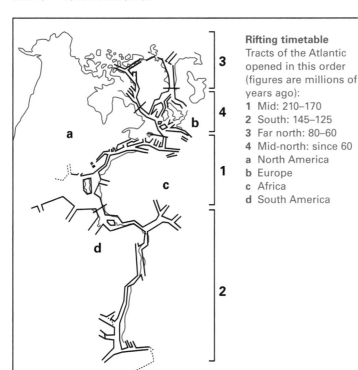

Rifting timetable
Tracts of the Atlantic opened in this order (figures are millions of years ago):
1 Mid: 210–170
2 South: 145–125
3 Far north: 80–60
4 Mid-north: since 60
a North America
b Europe
c Africa
d South America

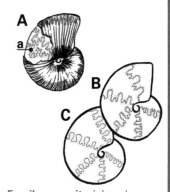

Fossil ammonite *(above)*
A Phylloceras—a Jurassic ammonite, an ammonoid with distinctive shell sutures (a)
B Ammonite shell (minus covering)—all sutures frilled
C Earlier ammonoid—not all sutures frilled

© DIAGRAM

11 CRETACEOUS PERIOD

Cretaceous world *(above)*
Lands might have looked like
this, omitting land bridges
and continental seas.

Cretaceous life *(above)*
1 Tyrannosaurus, an
   immense flesh-eating
   dinosaur
2 Magnolia, a flowering plant

The Mesozoic era closed with the Cretaceous period
(144–65 million years ago). Its name comes from the
Latin *creta*, meaning "chalk." Thick chalk deposits
formed in shallow seas that invaded Europe, North
America, and west Australia. Other deposits include
60 percent of today's known oil reserves.

Pangaea was now fragmenting into (northern)
Laurasia and (southern) Gondwana, and both of these
supercontinents were also cracking up.

Early on, a spreading rift opened up the South
Atlantic Ocean, driving South America and Africa
apart. Much later, rifting separated Scandinavia from
north Greenland, compressing and uplifting part of
Siberia to raise the Verkhoyansk Mountains. But land
still linked North America and Europe via south
Greenland and the British Isles.

Farther south, the Bay of Biscay gaped open as north
Spain pivoted away from west France. South of
Europe, Africa moved east, forcing "Adriatica" against
the Balkan plate, then moved west again. The
Mediterranean was forming as a pinched-off portion of
the Tethys Sea.

Dramatic changes added land and mountains to
North America. Early on, one part projected far toward
the North Pole. Then the north split open and the
north-west pivoted west, reacting with the Pacific plate
to ruck up rocks into the Brooks Range of north Alaska.
The Arctic islands also probably rotated to where they
lie today.

The overriding oceanic plates of western North
America continued spawning a great island arc of

Cretaceous cliffs *(right)*
Billions of shells of
microorganisms helped build
these chalk cliffs at Beer Cove
in southwest England.

Cretaceous life
These fossil organisms represent some of those that flourished in Cretaceous time. They are not shown to scale.

Plants
a Flowering plant
Invertebrates
b Mollusks
c Arthropod
d Echinoderm
Fishes
e Chondrichthyan
f Sarcopterygian

Reptiles
g Chelonians
h Plesiosaurs
i Choristodere
j Mosasaur
k Crocodilian
l Pterosaur
m Saurischian dinosaurs
n Ornithischian dinosaurs

Birds
o Ichthyornis
p Hesperornis
Mammals
q Insectivore

batholiths and Andean-type volcanoes. Late in the Cretaceous, North America's westward drift accelerated, speeding up subduction of Pacific crust. This crumpled up the Rockies, exposing metals manufactured deep down in the crust and realigned the rivers of the continent.

Meanwhile, in the Southern Hemisphere, India had cast adrift from East Africa, and New Zealand had most likely torn free from Australia.

Cretaceous climates remained chiefly warm or mild. Flowering plants began to spread. But the Mesozoic ended with the mass death of the dinosaurs and many other creatures. The impact of a huge asteroid might have caused the climatic changes leading to this mass extinction.

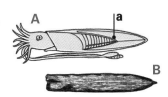

Belemnite (above)
A Reconstruction of a belemnite up to 31 inches (80 cm) long. A calcite rod (a), the guard, provided internal support for this Mesozoic kin of the squid and octopus.
B Fossilized belemnite guard. Fossil guards abound in some marine Mesozoic rocks.

© DIAGRAM

PALEOGENE PERIOD

Paleogene world *(above)*
The world looked roughly like this as India neared Asia in the Paleogene (alias early Tertiary) period.

This term is often used for the combined Paleocene, Eocene, and Oligocene epochs—the first part of the Cenozoic era. The Paleogene (65–23.8 million years ago) saw continents taking on their present shapes and locations, and birds and mammals filled the roles once taken by the dinosaurs. Spreading ocean floors and colliding and subducting plates raised mountains and reconfigured the map.

Shallow continental seas withdrew at first. Later, for a time the sea invaded parts of Africa, Australia, and Siberia.

Western North America was thrusting west and overriding cool oceanic crust. This crust warmed up deep down and expanded, lifting all of western North America about 30 million years ago. Meanwhile the Rocky Mountains and Colorado plateau were evolving. Volcanoes spewed ash, and lowland sediments formed vast oil-shale deposits. In the northeast, by 45 million years ago the widening Atlantic Ocean had parted North America from Europe. To the south, immensely thick sediments pushed the Mississippi Delta out into the Gulf of Mexico, and North and South America separated.

About 45 million years ago, Africa thrust north, driving lithospheric platelets into Europe. Island Iberia struck France and crumpled up the Pyrenees. Farther

Paleogene life *(right)*
1 Uintatherium, a rhinoceros-sized hoofed herbivore from Eocene North America
2 Gastornis, a giant ground bird of Eocene North America and Europe

east, the Adriatic plate overriding Europe's rim began pushing up the Alps. About 30 million years ago, part of France pivoted eastward, shoveling seabed sediments ashore on Italy to form the Apennines. Africa eventually added Sicily and the toe of Italy. Much of southeast Europe formed when two small Balkan plates struck southwest Russia.

In Africa itself great tracts warped up before splitting to release vast lava flows and open up the Red Sea rift.

By 40 million years ago, northward drifting India had struck Siberia and the small Kazakhstan and Tarim plates to its south. The impact concertinaed the collision zones and began to raise the Himalayas.

By 30 million years ago, Antarctica formed an Antarctic island continent. Chilled by its location, it indirectly helped to lower temperatures worldwide.

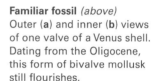

Familiar fossil *(above)*
Outer (**a**) and inner (**b**) views of one valve of a Venus shell. Dating from the Oligocene, this form of bivalve mollusk still flourishes.

Paleogene peak
Mountaineers scale a peak in the Rockies that was ice-sculpted as crustal heaving forced it high into the chilly upper air.

NEOGENE PERIOD

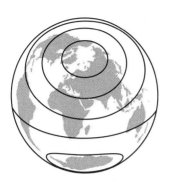

Neogene world *(above)*
Lands reached present positions by the end of the Neogene (late Tertiary) period.

The shrinking Tethys *(right)*
A map shows Miocene remnants of the once mighty Tethys Sea and (**a–c**) outlines of their modern relics, cut off by shifting crustal plates. The Mediterranean (**a**) once linked the Atlantic and Indian Oceans, and the Black Sea (**b**) joined the Caspian (**c**).

Neogene fossil *(above)*
Cypraea, the cowrie, is a mollusk whose oldest fossils crop up in rocks formed on the floors of Miocene seas.

The Neogene period (23.8–1.8 million years ago) comprises the combined Miocene and Pliocene epochs. Continents had almost reached their present places, and crashing plates were pushing up great modern mountain ranges.

By mid-Cenozoic times, subducting seabed was forging island arcs around the west and north Pacific Ocean. Pacific area plate movements also cast adrift whole strips of continental crust. Japan probably split away from mainland Asia. The peninsula of Baja California was torn from mainland Mexico, and rode northwest along with California west of the San Andreas Fault.

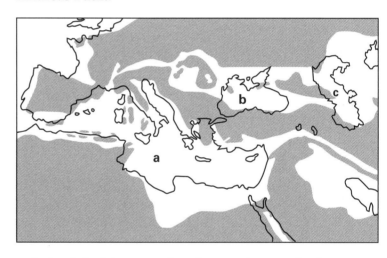

Indeed faulting or volcanic eruptions racked western North America from Alaska to Mexico. The Coast and Cascade Ranges sprouted. Fissure flows built a vast basalt plateau in Oregon, Washington, and Idaho. Block faults from Nevada to Mexico formed the parallel ranges and valleys of the Basin and Range Province. But the rising of the Rockies and Appalachians suggests an upwarping of the whole continent. Rejuvenated mountain rivers eroded sharply downwards; the Colorado River was now carving out the Grand Canyon. Farther south a land bridge rejoined South and North America, and volcanic peaks were rising in the Andes.

Out in the Pacific Ocean, volcanoes were spawning the Hawaiian Islands chain. Australia moving north collided with the Pacific plate. This forced up mountains in New Guinea and island "stepping stones" between Australia and Asia.

Meanwhile Africa's impact with Europe was manufacturing the Alps and the Carpathian and Atlas Mountains. The Red Sea rift prised Africa away from Arabia, and volcanic peaks arose along the African rift system. Arabia and Iran crashed into Asia to create the Taurus and Zagros Mountains, and advancing India built the Himalayas and Tibetan plateau.

By ten million years ago, Turkish and Arabian plates moving north had cut off the Mediterranean from the Indian Ocean, and Morocco had hit Spain. The isolated Mediterranean dried up and was refilled several times, leaving salt beds 20,000 feet (6,000 m) thick. Then, about five million years ago, a giant Atlantic waterfall burst in at Gibraltar and the sea refilled. Both polar regions now had ice caps.

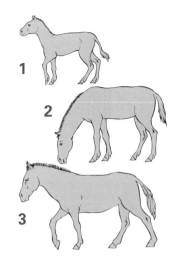

Evolving horses (above)
1 *Miohippus*, a small three-toed horse—early Miocene
2 *Merychippus*, larger and walking on each middle toe—Miocene
3 *Pilohippus*, the first one-toed horse—Pilocene

Rising land
A section through one side of the Grand Canyon (**a**) shows multilayered and faulted ancient rocks laid bare as the Colorado River (**b**) gnawed down through a rising plateau (**c**).

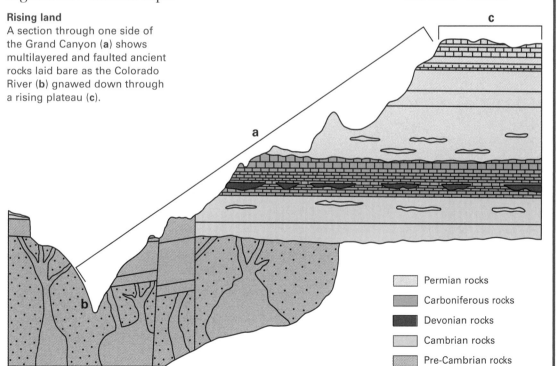

Permian rocks
Carboniferous rocks
Devonian rocks
Cambrian rocks
Pre-Cambrian rocks

© DIAGRAM

211

11 QUATERNARY PERIOD

Quaternary world *(above)*
Continents have shifted little in the last two million years, though levels of the land and sea have changed.

This followed the Tertiary period (also known as Paleogene and Neogene periods). The Quaternary began about two million years ago. It includes the Pleistocene or "Ice Age" epoch and the mild Holocene, or Recent, epoch in which we live today.

Pleistocene ice sheets smothered vast northern tracts, and glaciers filled mountain valleys worldwide. As the climate fluctuated ice repeatedly advanced and retreated—scouring valleys, damming lakes, rerouting rivers, and dumping debris over much of northern North America and Europe.

In intense glaciations sea level fell by as much as 330 feet (100 m). Meltwater torrents carved canyons in the rims of continental shelves. Land bridges joined Alaska and Siberia, mainland Asia and Indonesia, New Guinea and Australia, and the British Isles and mainland Europe.

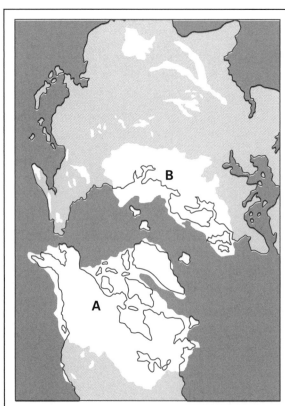

Lands under ice *(left)*
Pleistocene ice sheets sometimes covered these labeled parts of (**A**) North America and (**B**) Eurasia

Clues to cold
Below: This (much enlarged) foraminiferan coils right in warm water, left in cold.
Right: proportions of left- and right-coilers from 26 feet (8 m) seabed cores hint at past climatic changes.

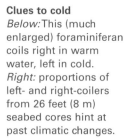

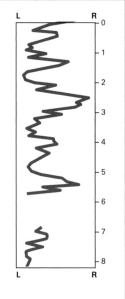

© DIAGRAM

Most of the northern ice sheets melted about 10,000 years ago. Sea levels rose, drowning the old canyons and land bridges. The Black Sea and Mediterranean were reunited. Along with the Caspian, both had once formed part of the prehistoric Tethys Sea until cut off by continental plates advancing from the south. As ice melted, crust once weighed down by ice bobbed up; it still rises in parts of Canada and Scandinavia. Meanwhile great subterranean forces were at work. Earthquakes, volcanic eruptions, oceanic trenches, and high peaks show where lithospheric plates still separate, collide, or grind against each other; shifting plates are still raising mountains and forcing oceanic crust into the mantle. Thus Cascades Range volcanoes sprout above the Farallon oceanic plate, which melts as it burrows under Oregon and Washington. And earthquakes shake the San Andreas Fault as the Pacific plate bears western California north.

Africa pushing under Europe forms a volcanic island arc in the Aegean and once raised Italy's volcanic Etna and Vesuvius. A rift broadens from the Dead Sea to the Gulf of Aden. Arabia thrusts against Iran. And farther east the Himalayas are still growing.

Humans evolved, but wildlife continues to wane as our inventive species competes with native plants and animals for food and living space.

Volcanic power *(right)*
In 1980 Mt. St. Helens proved explosively that a Cascades Range volcano can still be dangerously active.

11 TOMORROW'S WORLD

Future world *(above)*
Lands and oceans might be grouped like this 50 million years from now.

Short and long-term changes—some brought about by man—will drastically affect our planet's atmosphere, continents, and oceans.

By releasing chlorofluorocarbon sprays we have depleted the atmosphere's ozone and let in lethal quantities of ultraviolet radiation. By burning fossil fuels and felling forests we have warmed the atmosphere, shifting climatic zones and spreading deserts. But in a few thousand years, variations in Earth's tilt and distance from the Sun will cool the atmosphere, enlarging ice sheets and lowering the oceans. Later, continental drift will let warm ocean currents into polar regions. Then melting ice sheets may drown low land but cause a rebound of ice-laden Greenland and Antarctica.

Meanwhile, seafloor spreading and subduction will see oceans grow and shrink. In 30 million years the Atlantic will be much wider while the Pacific will have narrowed.

Dramatic changes will affect the land. Already human overuse of soil accelerates erosion and the accumulation of sediments in offshore waters. Longer term, mountains will be worn to stubs, while new ones rise as continents collide.

Sun, Earth, and ice
Changes in Earth's tilt and seasonal nearness to the Sun arguably produce ice ages.
A Earth far from the Sun in northern summer with Northern Hemisphere tilted slightly sunward: northern ice sheets grow.
B Earth close to the Sun in northern summer with Northern Hemisphere tilted far sunward: northern ice sheets melt.
a Earth in winter
b Earth in summer
c Sun

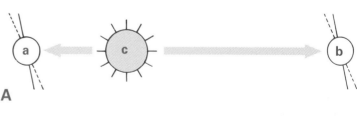

214

As Africa heads north, Europe's Alps may grow afresh, and the Rhine fault might split the continent. Iberia could be shoved into the Atlantic. The west Mediterranean may become a landlocked lake while squeezing destroys its east end and plants a mountain range from southeast Italy to Syria. Volcanic islands could appear offshore from Portugal to Norway.

Elsewhere, the Himalayas are still growing. In 15 million years, Australia might override Indonesia. In 25 million years, a north-south sea might split Africa in two. In 50 million years Los Angeles could be rafted north to join Alaska.

Billions of years will bring much more dramatic changes. A thinned asthenosphere may give Earth a rigid crust. Because our planet's spin is slowing down each Earth day could be more than 50 of today's days long. The Moon will move away and hover over the same part of Earth's surface, producing a fixed high tide. Eventually the Sun will swell up and engulf Earth. Later still, our solar system may fall into a dense black hole in the middle of our galaxy.

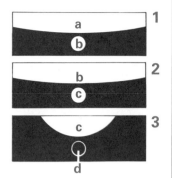

Black hole
Above: Diagrams stress the small size of a black hole.
1 Our Sun (**a**) compared to a white dwarf star (**b**)
2 White dwarf (**b**) compared to a neutron star (**c**)
3 Neutron star (**c**) compared to a black hole (**d**)
Below: A tiny but massive black hole (**a**) may distort the space-time fabric (**b**), creating an intense gravitational field that locks in electromagnetic radiation (**c**) and sucks stars in (**d**).

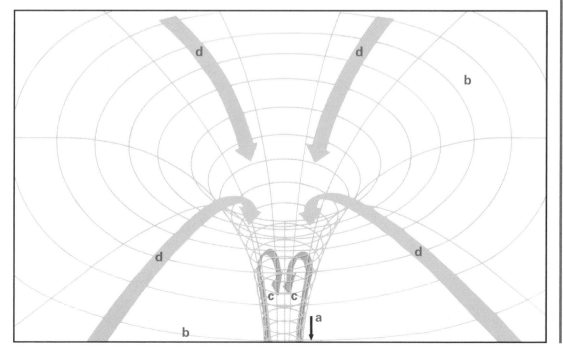

© DIAGRAM

CHAPTER 12 ROCKS AND MAN

In this chapter we briefly cover mapping rocks, collecting minerals, finding and extracting ores and other useful substances, and know-how used in putting rocks to work.

Two sixteenth-century woodcuts showing mining activities (First published in *De Re Metallica* by Georgius Agricola, 1556)

12 MAPPING ROCKS

+ Horizontal strata

30 ⟋ Inclined strata with degree of dip

⟋ Anticline

⟋ Syncline

30 ⟋ Minor fold, plunge in degrees

⟋ Base of lava flow, dotted above, base

⊥ Fault, with downthrow

Cu Mineral vein

Geological symbols (above)
Geologists use symbols such as these when mapping features of the rocks in any region.

Geological surveying
A field geologist takes a bearing by sighting a distant point through a mirror compass. Mirror compasses are particularly useful in poor light, even underground.

Mapping rocks involves detective work, for most solid rock (the "solid" geology) lies hidden under superficial deposits: residual remains of weathered rocks and "drift" laid down by rivers, glaciers, or wind. But solid rocks do show up here and there—in sea cliffs, river beds, quarries, road and railroad cuttings, boreholes, wells, pipeline trenches, and miners' spoil heaps. Even plowed fields may hold stones and boulders from the underlying rock. Elsewhere, the distribution of hills and valleys, springs, and even plants may indirectly hint at variations in the rocks below. Between them, such clues can indicate resistant and easily eroded rocks, faults, and layers of impermeable clay, or beds of limestone or sandstone.

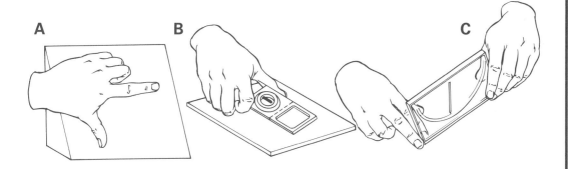

The field geologist visits rock exposures. He or she goes armed with at least a large-scale topographical map, compass, clinometer (or both combined in one), and rock-collecting tools. Between exposures, a hoe or hand auger may reveal rocks just below the soil. The geologist collects, identifies, and records rock and fossil samples, and plots exposures by compass on a map sheet. A clinometer will show dip (angle of tilt) of rock beds and plunge (tilt) of fold axes. A compass indicates strike (a horizontal compass direction at right angles to direction of dip). The mapmaker records angles of rock beds, faults, folds, joints, cleavage, and foliation, and notes features such as overturned beds, unconformities, intruded and extruded igneous rocks, and rocks metamorphosed by intrusions or other features.

Satellite imaging and aerial photography may aid observation in the field. Stereoscopic photographs stress features not always noticeable from the ground.

Once the geologist has recorded the types and angles of all exposed rocks he or she can then infer and map the hidden rocks between.

Strike and dip *(above)*
A The right-hand rule: when your thumb points down the dip, your index finger shows the bearing of the strike.
B Contact method for measuring strike: holding a compass horizontally, aligned parallel to strike, with compass edge against the rock surface or a map case laid on that to make it smooth and even.
C Using a clinometer to measure dip, at right angles to strike.

12 FINDING MINERALS AND FOSSILS

To find minerals and fossils, search rocks exposed by man or nature. Beaches, cliffs, gullies, rivers, road and railroad cuttings, gravel pits, quarries, and mine dumps may all prove fruitful.

Your first clues will come from guide books, geological maps, and local museums. Ask landowners' consent and avoid sites protected for their rarities, or cliffs liable to sudden rock falls. You should wear old clothes, strong shoes or boots, goggles, and perhaps a protective helmet.

Walk slowly, gazing on the ground. "Rock hounds" will seek freshly broken rocks, rock cavities, veins of calcite, and weathered fragments of attractive minerals. Fossil hunters search for exposed rocks containing bits of fossil plant or animal. Some rocks teem with fossils, others hold a few telltale shiny or discolored shapes, yet other rocks are sterile.

Finding minerals
An amateur "rock hound" wears a hard hat to hunt for geodes among rock debris fallen from a cliff containing shale and limestone layers. Only splitting open lumps will show what actually lies inside.

All major types of rock offer hope for people seeking decorative minerals. For instance, among igneous rocks, granites hold well-shaped crystals, and lava flows may include gas cavities rimmed with gleaming calcite or agate. Among sedimentary rocks, limestones and some shales, hold geodes—lumpy concretions often with crystal-lined internal cavities. Metamorphic rocks may feature colorful serpentine and marble. Many kinds of rocks hold veins of minerals. Even pebbles on a beach might be worth examining for agate, amethyst, and onyx.

Only sedimentary rocks hold fossils. The richest hunting grounds are marine limestones, shales, and certain sandstones. Likely finds are bits of fossil sea shell, coral, and hard parts of other animals without a backbone. But fossil land plants and animals crop up in certain rocks—the waste from coal mines for example.

1 Agate, a form of silica with parallel colored bands—often found in volcanic rocks

2 Granite, an igneous rock with coarse crystals of quartz, feldspar, and darker minerals

3 Marble, metamorphosed limestone that may be white, pink, black, green or streaky

4 Geode broken open to reveal decorative crystals that rim an internal cavity

5 Fossil ammonite embedded in a marine limestone rock

6 Fossil seed-fern leaf well preserved as a fine film in a rock slab from a coal mine

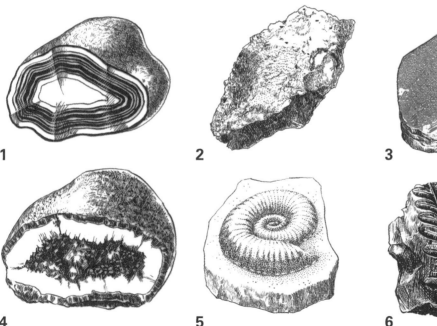

1

2

3

4

5

6

© DIAGRAM

12 EXTRACTING AND DISPLAYING FINDS

Finding a mineral or fossil specimen is just the start. Next you must free the specimen from its matrix (surrounding rock), transport it home, then clean, identify, and arrange it for display.

You can just pick up loose bits of rock. But use a geological hammer, a chisel, or a punch and awl to free mineral specimens embedded in a matrix. A pocket knife may come in handy, too. Try choosing pure specimens with undamaged crystals. Items should be big enough—but not too big—to put in a display. A shovel, sieve, rake, and gold pan may help you wash for heavy minerals in stream beds. A hand lens is useful for examining small specimens.

Rock hound's toolkit
1 Geological hammer
2 Chisel
3 Pocket knife
4 Gold pan
5 Hand lens
6 Cloth bag
7 Plastic bags
8 Notebook
9 Knapsack
10 Plastic goggles

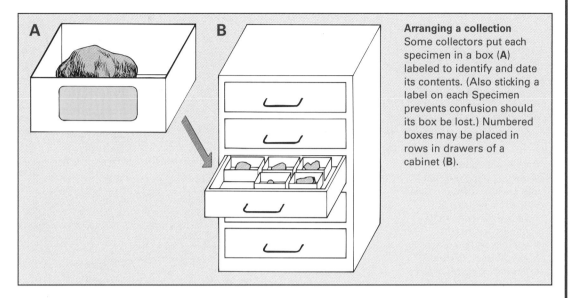

Arranging a collection
Some collectors put each specimen in a box (**A**) labeled to identify and date its contents. (Also sticking a label on each Specimen prevents confusion should its box be lost.) Numbered boxes may be placed in rows in drawers of a cabinet (**B**).

After collecting any item write its location in ballpoint on sticky tape and attach this to the specimen. Then wrap this in newspaper. If the piece is fragile, first wrap in tissue paper or cotton. Place all specimens from one locality in the same bag of doth or polythene, and label it.

Carry a notebook and pencil to record localities and mark these on a large-scale map.

Back home wash dirt and stains from specimens. Identifying most means choosing from 200 common compounds. With experience you can whittle down the possibilities by noting features such as color, hardness, luster, cleavage, and perhaps fluorescence or specific gravity. Field guide books, museum visits, and collecting trips with a museum class or rock and mineral club will all improve your diagnostic skills.

Arrange your mineral collection according to a numbering system. Paint a small white square on each item and write its number on the square in Indian ink. Then place items in an ordered display on shelves or in shallow drawers. Keep minerals dust free to secure the best effect.

Hardness scale
Softest (**1**) and hardest (**10**) minerals appear with others and everyday objects on a hardness scale devised by Austrian Friedrich Mohs in 1822. Each item scratches all those softer than itself.

1	Talc		
2	Gypsum		
3	Calcite	Fingernail	2½
4	Flourite	Silver	2½–3
5	Apatite	Teeth	5
6	Orthoclase	Penknife	5½
7	Quartz	Glass	6
8	Topaz		
9	Corundum		
10	Diamond		

© DIAGRAM

12 USEFUL MINERALS

Some field geologists seek ores—rocks rich enough in metals or certain other elements to be worth mining and separating from gangue (unwanted substances). Useful nonmetallic rocks and minerals include beds of gypsum, salt, and limestone. Igneous, metamorphic and sedimentary rocks all hold valuable substances, but only special areas reward a search.

Ores and magma *(right)*
A magma body injected into country rocks produces ores in several ways.
a Dense minerals settle in magma as it cools.
b Slow-cooling minerals crystallize in fissures.
c Magma minerals replace country rock in metamorphism
d Hydrothermal deposits of magma minerals form in fissures.

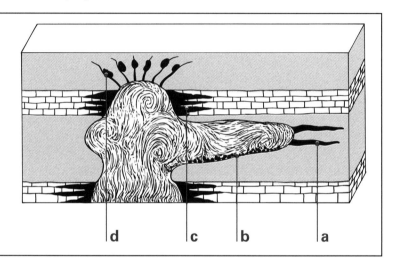

d   c   b   a

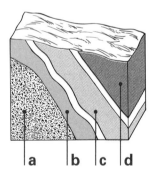

a   b   c   d

Ore sequence
A block diagram shows the mineralized zones produced as temperature and pressure changes affect mineral-rich solutions moving out from molten granite.
a Granite mass
b Tin deposits
c Copper deposits
d Lead and zinc deposits

Many ores form in or near a mass of molten magma, so igneous and nearby rocks are often fruitful hunting grounds. In slow-cooling basic magma, among the first minerals to crystallize and settle are apatite, magnetite, and chromite (respective sources of phosphorus, iron, and chromium). These ores occur as bands in sills and dikes.

Ores form, too, where a molten granite batholith injects hot, mineral-rich fluids under pressure into nearby rocks. As the fluids cool, their crystallizing minerals fill cracks, producing veins or groups of veins called lodes. Hot mineral-rich solutions losing heat and pressure as they filter out of granite replace nearby rocks with a sequence of tin, copper, lead, zinc, iron, gold, and mercury.

Certain useful substances occur as precipitates or evaporites. Thus hot springs precipitate the mercury ore cinnabar, and precipitation creates manganese nodules on the seabed. Evaporation of seas and inland lakes builds thick beds of gypsum, anhydrite, halite (rock salt) and potash.

Then there are residues or sediments. In the tropics, weathering breaks down aluminum silicate rocks yielding bauxite, the raw material for aluminum. Weathering also forms some ores of iron and

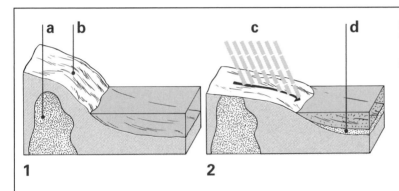

Sedimentary ores
1 Iron-rich pluton (**a**) inside a mountain (**b**)
2 Weathering and erosion (**c**) expose the pluton and wash dissolved iron into the sea. Precipitated iron oxide (**d**) accumulates in the seabed.

manganese. Percolating groundwater can deposit rich copper ores. Rivers and coastal waves sort particles of heavy minerals, dumping them on stream beds and beaches. Such placer deposits provide much of the world's tin, some gold and platinum, and diamonds and other gemstones.

Mineral reserves
This world map plots the land distribution of important metals and nonmetals.

● Metals
● Nonmetals

© DIAGRAM

12 | GEMSTONES

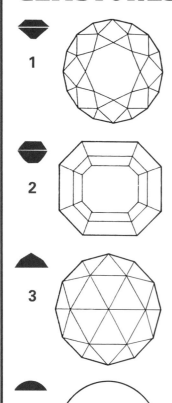

Major styles of cut
(Side views inset)
1 Brilliant
2 Step cut
3 Rose cut
4 Cabochon

Gemstones are minerals prized for beauty, durability, and rarity. Their worth depends on scarcity, color, purity, brilliance, hardness, and demand. Most stones are cut and polished so they glow and sparkle, then set in gold or otherwise as articles of jewelry.

Gemstones largely form in igneous or metamorphic rock from elements combining as they cool in gas pockets or superheated water. The coarse-grained igneous rock pegmatite is a source of beryl, tourmaline, and topaz. Volcanic andesite holds cavities where opal grew. Volcanic activity deep down produced the intense heat and pressure that forced the diamonds now found in pipes of kimberlite, a version of the rock peridotite. Emeralds are a transparent type of beryl that occurs in mica-schist, a metamorphic rock.

Jewelers identify gems by color, shape, hardness, refractive index ("light-bending power"), and specific gravity. Diamond, ruby, emerald, and sapphire—all transparent gemstones—rank with opaque opal as the precious stones. Semiprecious stones include agate amethyst, garnet, and peridot. Many gemstones are simply spectacular forms of ordinary-looking minerals. For instance, diamond is a pure, hard form of carbon— the substance coal is made of, and ruby and sapphire are just transparent, colored forms of corundum, much of which is drab or colorless. Metal oxides tint most gemstones, but diamonds get their color from a defect in their crystal structure.

Craftsmen cut and polish gems to bring out their special features. Translucent and opaque stones may get the rounded shape called cabochon. Transparent gems are faceted so that they reflect and bend the light. Each expert cuts a stone by grinding with a hard abrasive; diamond dust alone is hard enough to cut a diamond. Most stones are cut in one of four main styles, called brilliant, step, mixed (brilliant and step), and rose.

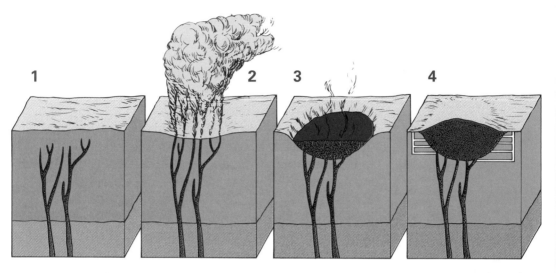

Diamonds' origins *(above)*

1 Intense heat and pressure create diamonds deep down in kimberlite pipes.

2 Gas exploding in fissures leaves a hollow in the land.

3 Kimberlite containing diamonds wells up and fills the hollow.

4 Miners sink shafts to reach the lower levels of the kimberlite.

The world's gems *(below)*
This map marks major sources of the five precious stones. Diamonds largely occur in old cratons but some were washed out as placer deposits.

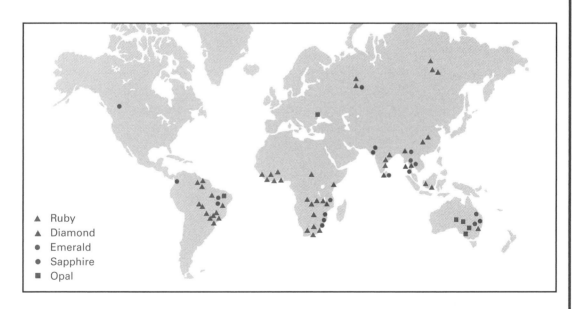

▲ Ruby
▲ Diamond
● Emerald
● Sapphire
■ Opal

© DIAGRAM

OIL AND GAS

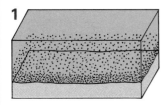

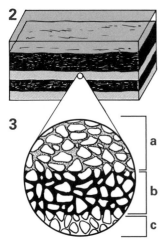

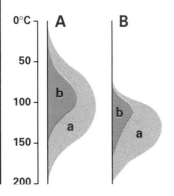

Hydrocarbons forming (*above*)
1 Dead organisms sink to the seabed.
2 Rocks cover them.
3 Bacterial action produces gas (**a**) and oil (**b**) above water (**c**) between sandstone particles (shown enlarged).

Coal, oil, and gas—the so-called fossil fuels—underpin industrial society today. All may have come from long-dead organisms that suffered incomplete decay. Coal is indisputably the carbon-rich remains of ancient forests. But the origins of oil and gas are more obscure and have been open to dispute

Most geologists believe that natural gas and oil derive from tiny marine organisms that died and sank to the seabed many millions of years ago. Compaction changed surrounding sediments to mudstones and shales. The resulting heat and pressure probably produced bacterial processes that helped transform the organisms into hydrocarbons—compounds mainly made of hydrogen and carbon.

Forces in Earth's crust supposedly drove most hydrocarbons from the rock in which they formed. They percolated up through permeable sand, sandstone, or limestone until trapped below a layer of impermeable rock. So the permeable rock below the trap became a fossil fuel reservoir. Here, natural gas floats on a layer of sticky to runny black to yellow liquid—a complex mix of hydrocarbons that we call petroleum. This petroleum floats on an even denser substance, water.

University astrophysicist Thomas Gold rejects the "squashed fish" theory. Gold argues that natural gas and oil originated in Earth's formation. He believes that enough of both fuels lies locked up deep in Earth to last us millions of years. In the late 1980s, a Swedish deep-drilling project with Gold as an adviser found gas traces in cavities 3.7 miles (6 km) below the surface. But more proof was needed to convince the skeptics. Most geologists still expect recoverable oil and gas supplies to dwindle sharply in a few decades.

Amounts and temperatures
Tests show relative amounts of gas (**a**) and oil (**b**) produced as temperatures increase with depth of burial.

A Amounts from marine plants
B Amounts from land plants

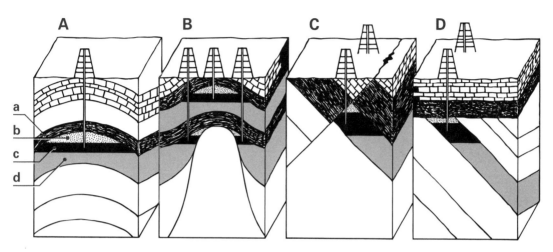

Reservoir rocks *(above)*
Here are wells drilled down through impermeable rock (**a**) and gas (**b**) to oil (**c**) trapped over water (**d**) in four types of situation.

A Oil in anticline
B Oil in rocks pushed up by a salt dome
C Oil trapped by fault
D Oil in tilted rocks trapped by unconformity

The world's oilfields *(below)*
Most of the oil-producing areas of the world lie in sedimentary basins. Oil does not survive igneous or metamorphic activity.

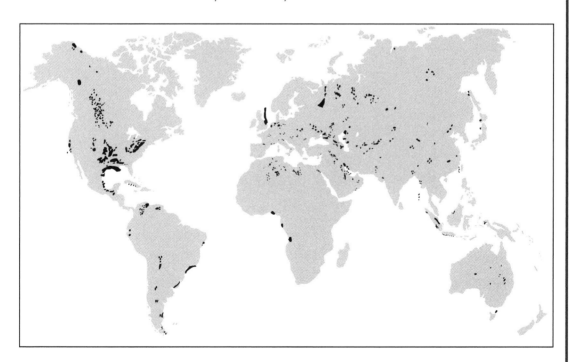

12 GEOLOGICAL PROSPECTING

To track down useful minerals or other substances below the ground the geologist gathers all information already known about the area to be explored. Next, he or she maps this geologically, noting surface clues like faults and gangue minerals including quartz. Then the geologist can bring to bear any of a battery of tests.

Gravimetric readings
A weight hangs from a coiled spring whose length varies with the force of gravity exerted by the rocks beneath.
a Normal reading
b High reading from dense igneous rock near the surface **c** Low reading from low density salt dome

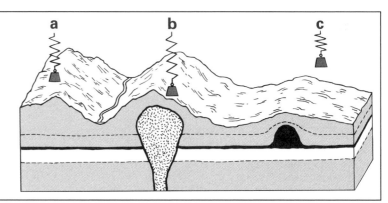

Geochemical tests analyze rock and other samples for trace elements that may lead the geologist to a major ore body.

Geophysical tests include the following. Geiger counters or scintillation counters detect radioactive substances such as uranium. Gravimeters reveal variations in the density and so the composition of underlying rocks.

Magnetometers indicate buried iron ores. Because iron is often found with sulfides, magnetometers may lead indirectly to nonferrous metals, too.

Magnetometer readings
(below)
Different rocks (**A**) produce local variations in Earth's magnetic field and so yield different readings (**B**) from a magnetometer.
a Country rock producing regional magnetism
b Topsoil producing background magnetism
c Deeply buried ores
d Ores just below the surface

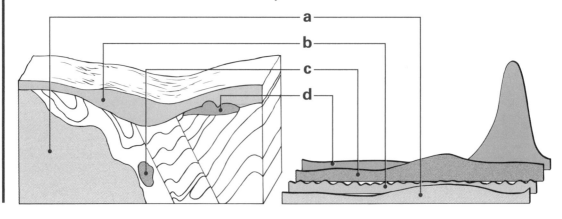

Electrical surveys show certain ores affecting natural ground currents related to Earth's magnetic field.

Seismic surveys test for various deposits, including oil, gas, and coal. Seismic surveying involves setting off explosions or vibrations that send shock waves down into the ground and timing their return from surfaces that bend or bounce them back. The speed of their return indicates the depth and nature of the rocks below.

All these prospecting methods merely hint at what lies underground. Only exploration can prove an ore is actually there and rich and big enough to be worth mining. If prospecting gives encouraging results, exploration follows. This means drilling sample cores or digging trial trenches to find out if development would pay. The next pages show just what development is likely to involve.

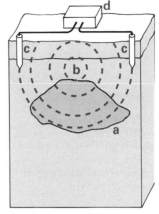

Electrical survey (above)
An ore deposit (**a**) affects natural ground currents (**b**) flowing between buried electrodes (**c**). A millivoltmeter (**d**) registers voltages at the electrodes.

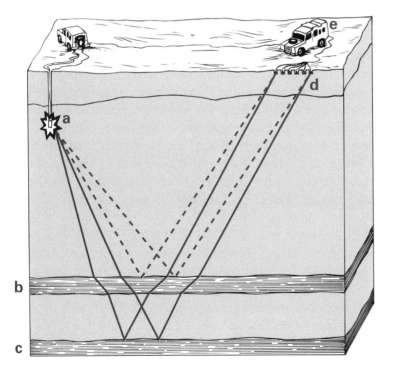

Seismic survey (left)
Dynamite exploded underground (**a**) produces shock waves, bent and bounced back by rock layers (**b**, **c**) at different depths to surface pickups (**d**). These relay the data to a recording truck (**e**).

© DIAGRAM

12 | MINING

Strip mining *(above)*
A power shovel (**a**) strips away superficial soil and rock called overburden (**b**) to expose a coal seam (**c**) just below the surface.

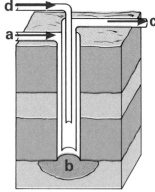

Pumping up sulfur
Hot water flowing down the outer pipe (**a**) of three concentric pipes melts underground sulfur (**b**) which rises through a middle pipe (**c**), forced up by compressed air from the inner pipe (**d**).

Some ores and useful nonmetallic minerals and rocks can be collected from the surface. Mining others may mean burrowing deep into Earth's crust.

Surface mining includes placer mining, dredging, strip mining, quarrying, and open-pit mining. Placer mining uses running water to sort heavy minerals such as tin, gold, and platinum from sand and gravel. Dredging employs mechanical scoops or buckets to raise sand or gravel from a pond or lake, before washing extracts heavy-mineral deposits. Strip mining strips off surface material so that power shovels can remove the underlying ore or coal. Quarrying uses saws, wedges, or explosives to free blocks of building stone from deposits near the surface, and power shovels to scoop up sand or gravel. Open-pit mining extracts ores from hard rock by cutting benches (steplike ledges) in a mountain.

Underground mining involves boring and blasting a hole into Earth's crust to reach an ore body. A nearly horizontal hole is called an adit, a vertical hole is a shaft. Horizontal passages following a vein are drifts. In level-and-shaft mining, a deep shaft leads to horizontal tunnels at different depths. These so-called levels lead to different parts of the ore body. They may be linked by sloping passages called raises and by holes called stopes produced by excavating ore. In room-and-pillar mining, miners excavate a large chamber to extract horizontal deposits of coal, lead, salt, or zinc. They leave rock walls and pillars to support the roof.

Pumping is another form of mining. Pumping sea water through precipitators extracts magnesium from sea water. Underground salt and sulfur can be pumped up in liquid form through boreholes. Boreholes drilled on land and on continental shelves also tap rock reservoirs of oil and gas.

All mining processes call for an understanding of the strengths and weaknesses of rocks and the special problems posed by faults and folds.

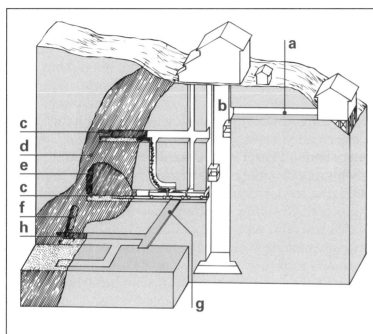

Underground mine *(left)*
Most items depict those named in the text.
a Adit
b Shaft
c Levels
d Ore body
e Stope
f Raise
g Winze (driven down from a level)
h Drift

Room and pillar mining *(below)*
This view stresses coal pillars left by first workings, but omits most of the rock roof they support.
a Unworked coal seam
b Coal pillars
c Rooms left by excavated coal
d Rock roof
e Coal conveyor
f Coal stockpile

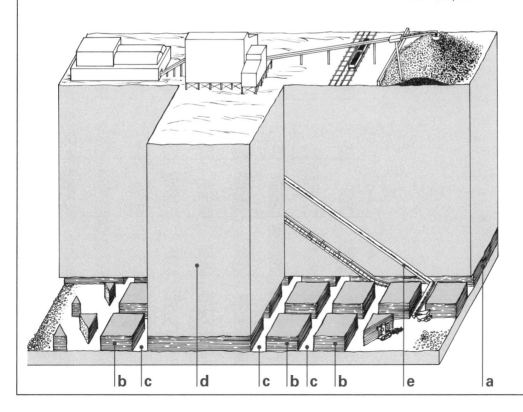

12 | WORKING WITH ROCKS

Geological know-how plays a key role in modern mining and other major works involving excavation. For without such scientific information engineers might make dangerous and costly errors that burst dams, topple buildings, or let tunnel roofs collapse.

Engineering geologists study such properties of rocks as hardness, toughness, elasticity, durability, and permeability. These depend largely on mineral ingredients. Thus quartz is harder than steel and chemically durable, while mica is much softer than steel and its dark variety biotite is readily attacked by rainwater. Grinding, crushing, and other tests hint at a rock's behavior when excavation or building reduces or increases the pressure on it, and perhaps exposes it to frost and moisture.

Similarly soil mechanics deals with properties of clays, silts, sands, gravels, organic soils, and peat— their capacities to carry loads, and liabilities to settlement, compaction, permeability, and frost.

Properties of rocks
Bar heights indicate actual (**1–2**) and relative (**3–6**) values for 10 types of rock used in roads and concrete. In **1–4** the higher the bar the higher the quality; in **5–6** the reverse.

Properties:
1 Specific gravity
2 Crushing strength
3 Hardness (abrasion)
4 Toughness (impact)
5 Abrasion value-dry (dry attrition value)
6 Abrasion value-wet (wet attrition value)

Rocks:
A Basalt
B Flint
C Gabbro
D Granite
E Gritstone
F Hornfels
G Limestone
H Porphyry
I Quartzite
J Schist

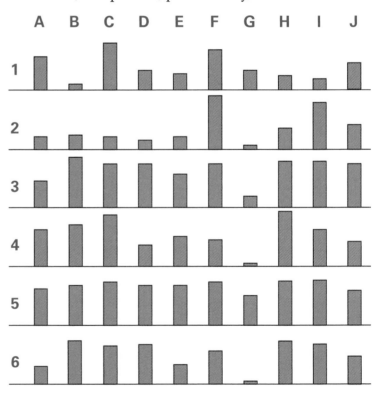

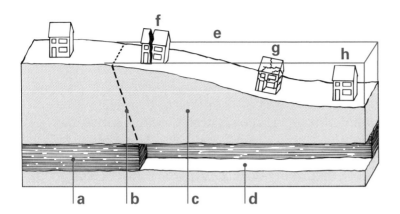

Mining subsidence *(left)*
This diagram shows three possible effects on buildings.
a Unworked coal seam
b Worked coal seam
c Subsiding rock
d Limit of subsidence
e Presubsidence land level
f Tension damage
g Compression damage
h Vertical displacement

Then, too, geological mapping of a chosen area reveals local weaknesses like faults/joints, bedding planes, landslides, and soil-filled river channels snaking through a bed of solid rock.

All this information helps to show where engineers can safely carve out tunnels, cuttings, quarries, and canals, or excavate foundations for dams, bridges, and large buildings. Geological studies also show sites best avoided or requiring special structures. Thus engineers can throw a slim, concrete arch dam across a narrow gorge walled by strong, elastic rock. But such a dam would burst if built on soft, weak shale. Here, they would raise a massive earth- and rock-filled dam instead. Similarly, New York's resistant metamorphic rock provides a firmer base for skyscrapers than the soft clay London stands on.

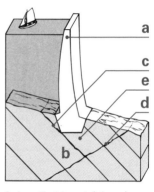

A dam that burst *(above)*
a Concrete arch dam
b Bedrock
c Crack along foliation plane
d Crushed seam
e Rock wedge
Water seeping under pressure into (c) dislodged the wedge and burst the dam.

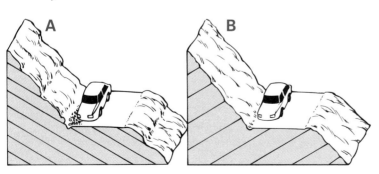

Safe and unsafe *(left)*
Two hillside road cuts
A Unsafe. Dip of bedding may produce collapse.
B Safe. Dip of bedding protects against collapse.

MAN-MADE ROCKS

Man-made landscape
City skyscrapers, streets, and other structures are made from rocks extracted from the ground and reconstructed into concrete, steel, glass, bricks and other man-made substances.

Rocks and the landforms made by their erosion helped decide where people built towns, ports, and cities. Now, the environmental geologist works with rocks to find supplies of water and building materials, and to check building-site stability, assess local risks of flood or earthquake, trace the origins of toxic chemicals and polluted water, and choose safe dumps for refuse burial.

The immense extent of modern buildings, airports, roads, and railroads means that these in turn affect the surface of the continents. Across mighty tracts of land,

workers have masked the native soil with man-made rocks. Their chief materials are clay bricks and tiles; walls and floors of concrete (sand, stones, lime and other substances bound by water); roads of stones set in bitumen; glass windows of silica, lime, and soda; and steel rods and girders of iron with added elements.

Most construction works affect the natural processes that wear away and build dry land.

Some projects curb erosion: dams raise the base to which rivers erode their beds; groynes and sea walls hamper the destructive work of waves. By drying swamps and marshes, drainage actually speeds up land formation. Reversing the erosion process altogether, offshore dredging scoops building sand and gravel from the continental shelf, and Dutch engineers have won much of the Netherlands from the sea.

But human handiwork accelerates erosion, too—by mining and quarrying, and through farming malpractice that loosens soil until rain or wind wash or blow the soil away. Also, miners extract fossil fuels and certain minerals far faster than Earth replaces them.

For ill or good, our species is a geologic agent strong enough to tamper with our planet's crust.

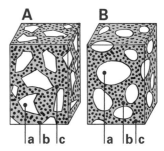

Man-made rock *(above)*
Concrete is a man-made equivalent of conglomerate.
A Concrete
B Conglomerate
a Clasts
b Matrix separating clasts
c Cement bonding clasts

Man-made land *(below)*
A Area of the Netherlands above sea level
B Actual land area today, increased by swamp drainage and reclamation from the sea. Gray areas show land below sea level.

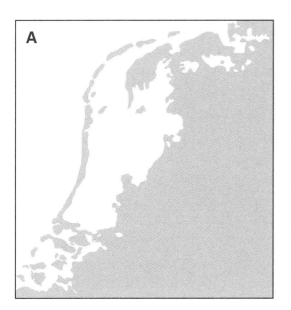

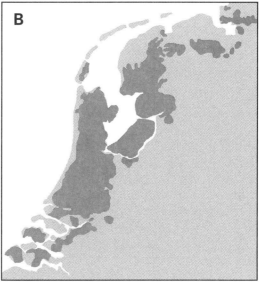

CHAPTER 13 MONITORING EARTH

Ice sheets flow and continents drift across the surface of the globe unimaginably slowly. Mountain ranges hundreds of miles long can take a thousand years to rise or subside an inch. In the past, scientists could not measure the tiny movements of such vast geologic features using conventional instruments. Satellites and remote sensing technologies have changed all this. Unblinking artificial eyes now scan every square foot of Earth's surface from orbit. Instruments that can "see" the chemical composition of rocks from a thousand miles in the sky now circle the globe every 90 minutes. This chapter looks at the way these new tools are revolutionizing Earth science.

Images of Earth produced by orbiting satellites:
1 Andes Mountains, Chile
2 College Fjord, Prince William Sound, Alaska
3 Mt. Rainier, Washington
4 U.S.–Mexican border between Mexicali and Calexico
5 Nasca Lines, Peru
6 Thar Desert, Northwest India

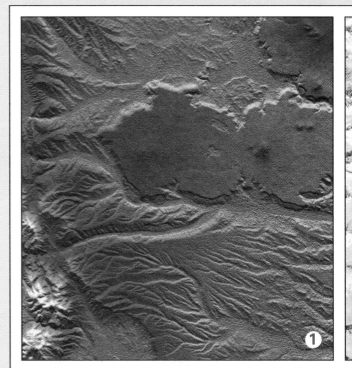

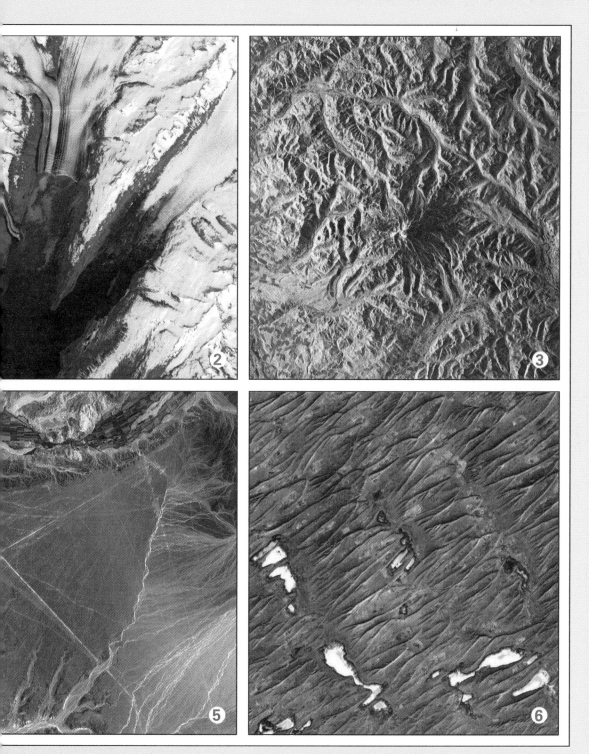

13 REMOTE SENSING 1

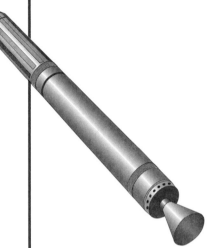

Explorer 1
In January 1958 the United States launched its first satellite, *Explorer 1*. Although preceded by the USSR's *Sputnik 1* a few weeks earlier, *Explorer 1* was the first satellite to make a scientific discovery. This tiny vehicle, less than 40 inches (100 cm) long, gathered data that led to the discovery of Earth's Van Allen radiation belts. It was the world's first Earth observation satellite.

Satellites have allowed scientists to study our planet in new ways. Satellites can scan the whole Earth in a few days collecting data that would take scientists on the ground decades to gather. Collecting information in this way is known as remote sensing. This means that the information is collected "at a distance" rather than by physical contact. Remote sensing is also carried out by aircraft and ships.

Radiometry is a form of remote sensing. It is defined as the measurement or recording of electromagnetic radiation. This includes radio waves, and visible, infrared, or ultraviolet light.

Passive radiometry is the measurement of electromagnetic radiation given off by, or reflected, naturally by an object. For example, when scientists look at the Moon with a telescope they are performing passive radiometry because they are collecting sunlight that has been reflected by the Moon. Active radiometry involves emitting electromagnetic radiation and sensing any that is reflected back. It is similar to using a flashlight to look into a dark cave. Radar is an example of active radiometry because it works by emitting a pulse of microwave radiation and then sensing any that is reflected back.

Satellites that are used to study Earth are often referred to as remote sensing satellites or as Earth observation satellites. They use passive or active radiometry. Active radiometry has the advantage that it can be used through cloud and over the night side of the globe. For example, satellites that use radar to map surface features can scan areas that are in total darkness or under thick cloud just as easily as they can scan sunlit areas under clear skies. Passive radiometry usually collects sunlight reflected from the surface, so it does not work over areas that are in darkness or obscured by weather. On the other hand, much more detailed information can usually be collected by passive instruments that scan reflected sunlight because much more radiation is available.

Radiometry

Radiometry is the detection or measurement of electromagnetic radiation. Looking at a rockface is a form of radiometry. Visible light (a form of electromagnetic radiation) is reflected by the rock and collected by your eye (a radiometric sensor). The brain then forms an image from the collected data.

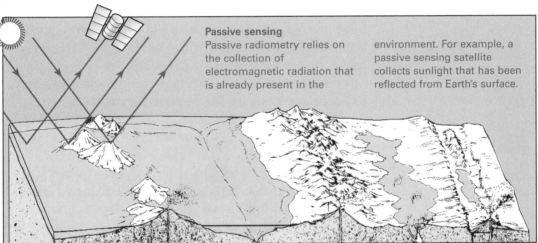

Passive sensing

Passive radiometry relies on the collection of electromagnetic radiation that is already present in the environment. For example, a passive sensing satellite collects sunlight that has been reflected from Earth's surface.

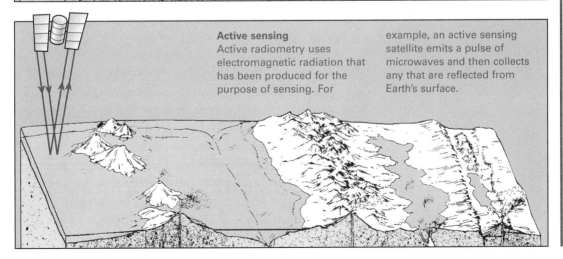

Active sensing

Active radiometry uses electromagnetic radiation that has been produced for the purpose of sensing. For example, an active sensing satellite emits a pulse of microwaves and then collects any that are reflected from Earth's surface.

13 | REMOTE SENSING 2

Visible areas

A satellite in a high Earth orbit can scan almost one half of the surface at the same time. A satellite in a low Earth orbit with the same field of view can only scan a comparatively small area.

x°

A Area visible to GEO satellite (altitude 22,254 miles, 35,800 km)

B Area visible to low orbit satellite (altitude 438 miles, 705 km)

x° Same viewing angle for both

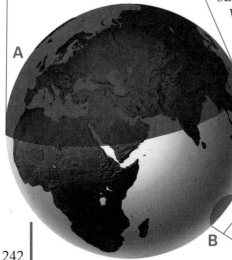

A

B **x°**

Remote sensing satellites and other spacecraft associated with Earth sciences orbit the planet in different ways. There are two important elements to consider—the orbit's inclination and its altitude.

Altitude simply refers to the distance of the orbit above Earth's surface. The lower a satellite's orbit, the faster it will complete each circuit of the planet. The higher a satellite's orbit the more of the surface it can "see" at the same time.

Inclination is the angle of the orbit with reference to Earth's equator. An orbit that follows or is exactly parallel to the equator is said to have an inclination of zero degrees to the equator. An orbit that passes over the poles is at right angles to the equator and is said to have an inclination of 90 degrees to the equator. Orbits with inclinations of 90 degrees or close to 90 degrees are commonly known as polar orbits.

Remote sensing satellites that are intended to carry out repeated scans of the planet every few days are usually placed in polar orbits at relatively low altitudes. A satellite in low polar orbit will circle Earth in about 90 minutes. As the satellite returns to the starting point of its orbit, Earth will have rotated so that the next orbit will sweep over a strip, or swath, of the surface slightly to the west of the first. With each successive orbit the satellite will pass over a new area. Within a few days the vehicle will have scanned the entire Earth in a sequence of overlapping swaths.

Some satellites are placed in Sun-synchronous polar orbits. This means that they cross the equator at about the same time of day on every pass. A Sun-synchronous orbit allows the satellite's instruments to view the same latitudes at the same time of day on each successive pass. This is important for passive sensing instruments because it means that the areas they are viewing are illuminated by roughly the same amount of sunlight coming from the same angle each time. Changes over time are much easier to spot under these conditions.

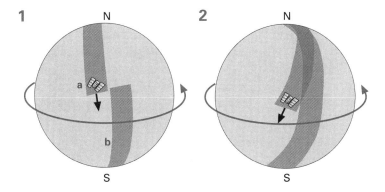

For other satellites, it is more important that the instruments are in the line of sight of large areas of Earth's surface for long periods than that all areas are scanned for relatively short periods. Weather and communication satellites are commonly placed in high-altitude orbits that have these characteristics.

Swaths and orbits

Remote sensing satellites that observe from low polar orbit scan a strip of the surface on each orbit. This strip is known as a swath. As Earth rotates beneath the satellite the swath is displaced to the west. With each orbit the scan swath is wrapped around the planet like a band of ribbon around a ball. Eventually, the whole surface is scanned.

1 A satellite (**a**) scans the surface beneath it continuously. The strip of Earth that it scans is known as a swath (**b**).

2 Since Earth is rotating as the satellite orbits, the swath covers more and more of the planet until the whole surface has been scanned.

Geostationary orbits

A satellite in geostationary orbit (GSO) makes a circular orbit with an altitude of 22,245 miles (35,800 km) and travels in the same direction as Earth's rotation. At that altitude it takes exactly 24 hours to complete one orbit. A satellite in such an orbit remains in a fixed position relative to a point on Earth's surface. Weather satellites are usually placed in GSO. The image shows how just four satellites in GSO can provide coverage of the entire globe. *Geostationary Environmental Operational Satellite* (*GOES*) *9*, *10*, and *12* are U.S. satellites. *Meteosat 6* is European.

GOES 12

GOES 10

Meteosat 6

GOES 9

© DIAGRAM

13 LANDSAT

The Landsat program has been operating for more than 30 years. It is the world's longest running satellite Earth observation program and has been a cooperative effort by the National Oceanic and Atmospheric Administration (NOAA) and NASA. The first Landsat satellite (*Landsat 1*) was launched in 1972 and remained operational until 1978. The seventh, *Landsat 7*, was launched in 1999. The millions of images of Earth and other data collected by Landsat instruments through this period are a very important resource for scientists engaged in monitoring changes across the planet. Earth monitoring is only useful when there is a consistent record of data over time. The Landsat spacecraft are a good example of the way that Earth observation satellites have developed over the years.

Landsat 1 flew in a polar orbit at an altitude of 564 miles (907 km), traveling once around Earth every 103 minutes. It carried two primary instruments, a video camera for taking visible light and infrared photographs, and a multispectral scanner for collecting data in four bandwidths of the visible and infrared

Swaths
The swath scanned by *Landsat 7* on each orbit is 114 miles (185 km) wide (**a**). Each swath is 1,710 miles (2,752 km) (**b**) to the west of the previous swath. After 16 days, the scanned swaths will have covered the entire globe. The first swath of the 17th day follows the same track as the first swath of day one.

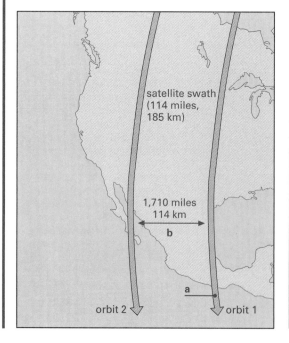

satellite swath
(114 miles,
185 km)

1,710 miles
114 km
b

a

orbit 2   orbit 1

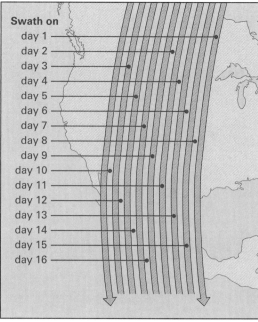

Swath on
day 1
day 2
day 3
day 4
day 5
day 6
day 7
day 8
day 9
day 10
day 11
day 12
day 13
day 14
day 15
day 16

spectrum at the same time. The multispectral scanner collected data that allowed scientists to "read" the characteristics of the area being scanned. It had a maximum resolution of 260 feet (80 m), meaning that each pixel of the images it produced corresponded to a square on the ground with sides 260 feet (80 m) long. Video recording devices on the satellite could store up to 30 minutes of images from the instruments. *Landsat 1* proved the effectiveness of multispectral scanners for Earth observation.

Landsat 7, launched 27 years later, is equipped with much more sophisticated multispectral scanners. It scans simultaneously in four visible and near infrared bandwidths, two shortwave infrared bandwidths, and one thermal infrared bandwidth. Some of its instruments have maximum resolutions of 98 feet (30 m), others of just 16 feet (5 m). This is comparable to being able to see features the size of a supermarket using *Landsat 1* instruments and features the size of a boxing ring with *Landsat 7* instruments.

Multispectral scans
The multispectral scanners on *Landsat 7* are a typical example of modern passive radiometry sensors. This image of the port city of Kobe, Japan shows how urban areas dominated by concrete surfaces (colored areas) are visible against rural areas dominated by vegetation (gray areas). Landsat instruments are capable of distinguishing many different types of terrain.

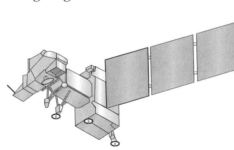

Landsat 7
NASA's *Landsat 7* Earth observation satellite operates from a Sun-synchronous polar orbit at an altitude of 438 miles (705 km). The satellite takes 99 minutes to complete one orbit and can scan all of Earth's surface between 81°N and 81°S every 16 days. It crosses the equator traveling north to south at between 10:00 a.m. and 10:15 a.m. local time on each orbit.

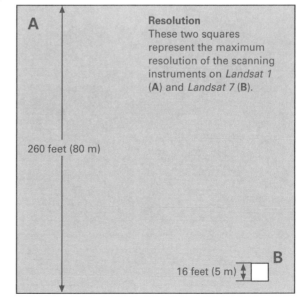

A

260 feet (80 m)

Resolution
These two squares represent the maximum resolution of the scanning instruments on *Landsat 1* (**A**) and *Landsat 7* (**B**).

16 feet (5 m)

B

© DIAGRAM

GLOBAL POSITIONING SYSTEM

The Global Positioning System (GPS) was one of the most revolutionary and important innovations of the twentieth century. It has had wide-ranging applications in many fields and is critically important to many forms of modern Earth science. GPS provides precise data about time and position. The most sophisticated GPS equipment can give a three-dimensional position accurate to within 0.4 inches (1 cm) and a time check accurate to within one microsecond. Many remote sensing instruments mounted on satellites rely on having precise positional and time data and most receive this data from the GPS. Static ground-based instruments also commonly use the GPS time signal.

GPS consists of a group (or constellation) of 24 satellites in circular Earth orbits at an altitude of 12,600 miles (20,200 km). Each satellite completes two orbits every day and they are spaced apart in such a way that at least four satellites are

GRACE
Two identical satellites flying about 137 miles (220 km) apart use the GPS to measure changes in their position relative to each other.
a Grace satellite
b Grace satellite
c GPS satellites

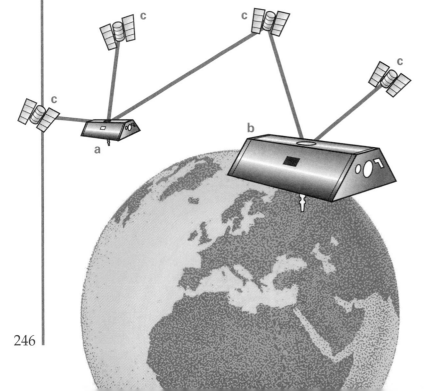

NAVSTAR GPS *(above)*
The satellites of the Navigation Signal Timing and Ranging Global Positioning System (NAVSTAR GPS) are operated by the U.S. Air Force Space Command. More than 50 have been launched since 1978. Twenty-four are active at all times. Six groups of four satellites orbit in six different orbital planes.

always within the line of sight of any point on the globe. Each satellite transmits a time signal and its orbital position constantly. A GPS receiver can calculate its position relative to the four nearest GPS satellites using this information.

NASA's Gravity Recovery and Climate Experiment (GRACE) is a good example of a satellite system that relies on GPS. GRACE consists of two identical satellites flying in close formation. Both satellites take constant measurements of their positions relative to each other and the ground as they study Earth's gravity field. Positional data for these measurements comes direct from the GPS.

The Plate Boundary Observatory program (PBO) is an example of a ground-based system that uses GPS for Earth science. PBO utilizes more than 1,000 GPS receivers scattered across the United States to monitor movements along plate boundaries.

Comparing delays
A GPS receiver (**a**) picks up time signal transmissions from four GPS satellites (**b**). All four satellites have synchronized clocks. The signal from each satellite takes a different period of time to reach the receiver depending on the distance they have to travel (D1 to D4). The differences between the time signals provides the information needed for the receiver to calculate its position relative to the satellites.

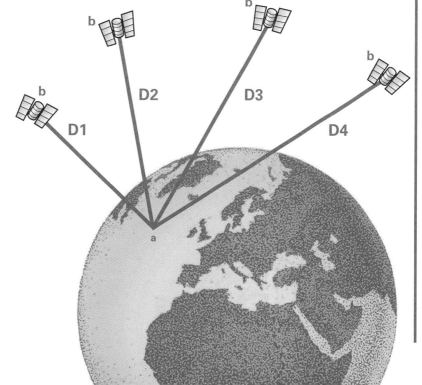

© DIAGRAM

13 MAPPING THE SURFACE

Accurately mapping Earth's surface has always been a challenge. Modern remote sensing instruments have now made it possible to map most of the globe with a high degree of accuracy. Spacecraft carrying these instruments are not only able to measure the relief of the land, but can also sense whether they are looking at a surface covered by vegetation, ice, or simply bare rock. They also have the advantage of being able to survey remote areas that would be inaccessible to ground-based surveyors.

In 2000, NASA's Space Shuttle carried radar equipment into orbit that was used to make a high-definition topographical model of Earth's surface. The Shuttle Radar Topography Mission (SRTM) used a synthetic aperture radar to create three-dimensional images of about 80 percent of the world's landmass with a margin of error between zero and 52 feet (0–16 m). This was the most accurate set of global topographical data ever produced. The model created by the SRTM has been used in a wide range of applications including improved flood control planning, volcano monitoring, earthquake research, and glacier and ice sheet monitoring.

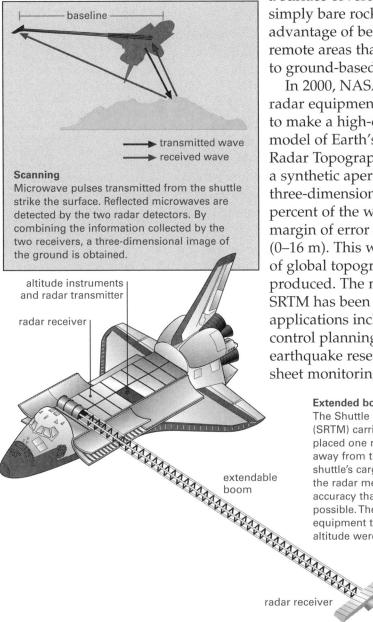

baseline

→ transmitted wave
→ received wave

Scanning
Microwave pulses transmitted from the shuttle strike the surface. Reflected microwaves are detected by the two radar detectors. By combining the information collected by the two receivers, a three-dimensional image of the ground is obtained.

altitude instruments and radar transmitter

radar receiver

extendable boom

radar receiver

Extended boom
The Shuttle Radar Topography Mission (SRTM) carried an extendable boom that placed one radar receiver 195 feet (60 m) away from the other radar receiver in the shuttle's cargo bay. This long "baseline" gave the radar measurements much greater accuracy than would otherwise have been possible. The radar transmitter and equipment to measure the shuttle's exact altitude were also in the cargo bay.

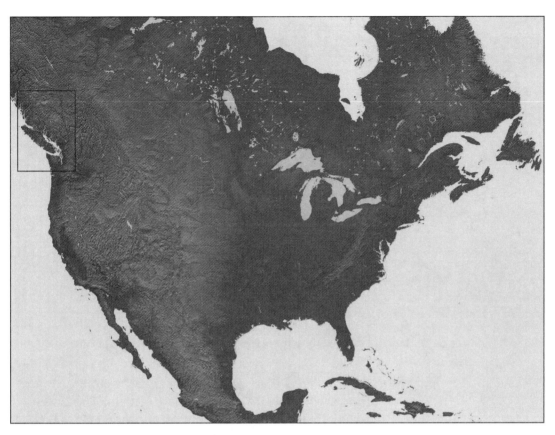

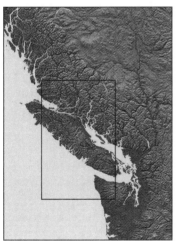

Zooming in
Before the SRTM, the smallest details on the best global topographical data were 0.4 square miles (1 km^2). SRTM provided a model in which the smallest details are 323 square feet (30 m^2).

ICE COVER

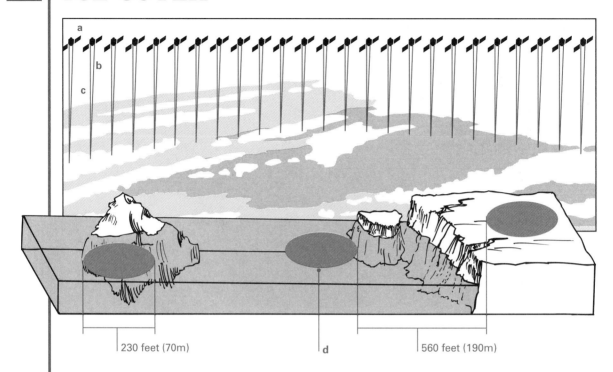

230 feet (70m)

d

560 feet (190m)

GLAS

ICESat carries a lidar instrument known as the Geoscience Laser Altimeter System (GLAS). Pulses of laser light are fired at the ground 40 times a second and the reflected light is detected. The laser strikes patches of ground that are about 230 feet (70 m) in diameter and about 560 feet (170 m) apart (known as sensor footprints). GLAS can also fire laser light that is reflected by clouds and particles in the atmosphere.

a *ICESat*
b laser pulse
c reflected laser pulse
d sensor footprint

The ice sheets that cover Earth's polar regions are very important elements of the climate system. These vast frozen areas are far from being static, however. They grow, shrink, and flow in ways that climate scientists are still trying to understand. It is known that ice sheets have had a strong influence on sea levels in the past and there is concern that global warming may lead to rapid changes in the ice sheets in the near future.

There are several satellites studying the polar regions with remote sensing radiometric instruments and more are planned. 1n 1997, the Canadian Space Agency satellite *Radarsat* made the first high-definition radar map of the whole of Antarctica. This data provided a framework for the observations made by NASA's *Ice, Cloud, and land Elevation Satellite* (*ICESat*) launched in 2003. *ICESat* used a lidar instrument to collect highly accurate topographical data from across Antarctica. The same instrument also collected data about the distribution and makeup of clouds over the continent. A European Space Agency satellite named *Cryosat II* is due to be launched in 2009. It will provide complete topographical maps of Antarctica every month.

Climate scientists want to find out about the topography of the ice sheets because this can help them understand the balance between the formation of new ice and the melting of existing ice. Ice sheets are in constant motion. At the edge of Antarctica's ice sheet, along the coast, some ice is always melting or breaking off to form icebergs. The material that is lost in this way is replaced by newly-formed ice flowing down to the coast from the colder interior of the continent. Some scientists believe that a warming climate may not result in less ice overall. Higher temperatures will cause more ice to melt along the edges of the continent, they argue, but it may also result in more moisture in the atmosphere which may in turn result in greater snowfall across the interior. This increased snowfall would result in a more rapid formation of ice replacing that lost at the coast.

Factors influencing ice formation
a Precipitation and evaporation
b Atmospheric circulation
c Snowfall
d Sublimation
e Ocean temperature and circulation
f Ice flow
g Accumulation and wind redistribution
h Melt and run off

Retreating ice
Constant monitoring of the north polar region has revealed a significant reduction in the amount of summer sea ice over a period of 25 years.

September 1980

☐ sea ice

September 2004

Arctic Circle

Ice sheet uncertainty *(left)*
Climate scientists are unsure about the effect of rising temperatures on the world's ice sheets. Ice melts at the edges and is replenished by ice flowing from the center. A warmer climate may increase evaporation from the oceans and encourage precipitation over the interior. Other variables include changes in sublimation from the surface of ice sheets and the redistribution of snow by changing wind patterns.

© DIAGRAM

13 | VOLCANOES

Hot spots *(below)*
The Nyiragongo volcano in the Democratic Republic of Congo erupted in January 2002. This image shows a shaded relief map of the region created from Shuttle Topographical Mapping Mission data along with hotspots picked up by MODIS less than an hour after the eruption occurred. Each square is a 0.4 square mile (1 km²) area. They show where lava flowed from the volcano into the town of Goma.

There are at least 1,500 potentially active volcanoes on Earth. Many of the largest and most active are monitored 365 days a year by ground-based instruments, but not all of them can be watched in this way. Satellite remote sensing technology has allowed volcanologists to detect newly erupting volcanoes in remote locations quickly. For example, in October 2001 Mount Belinda on the South Sandwich Islands began erupting for the first time in recorded history. Thanks to satellite data, scientists were aware of the event within hours. Without satellites an eruption in such a remote location may have gone unreported for months.

Studying active volcanoes is important because it gives scientists the opportunity to gather data about the changes that a volcano goes through before and during an eruption. This improves both their ability to predict the behavior of volcanoes that may threaten human habitations and their understanding of Earth's geology. Satellites are also used to study the flow and cooling of lava, the spread of volcanic ash through the atmosphere, and the release of volcanic gases.

Myamuragira

Nyiragongo

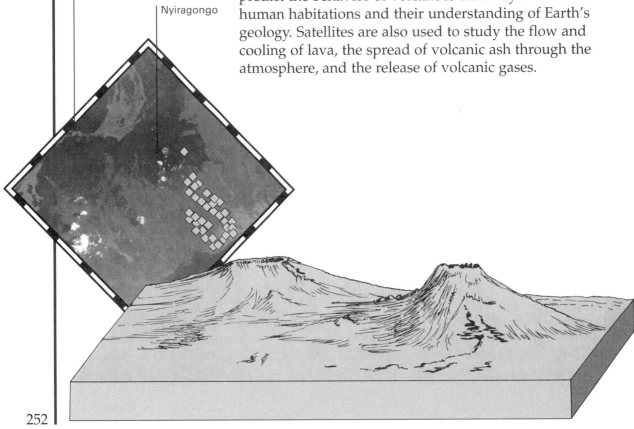

Volcanologists make use of both passive and active radiometric instruments. For example, to locate heat sources created by new eruptions scientists make use of data from the Moderate Resolution Imaging Spectroradiometer (MODIS). A MODIS instrument is carried on two of NASA's most advanced Earth monitoring satellites *Terra* and *Aqua*. Both of these satellites scan every spot on Earth's surface once every two days.

Before a volcano erupts the sides of the mountain may start to bulge from the pressure building up beneath the surface. Volcanologists look for these deformations. Accurate topographical maps produced from remote sensing data make it possible to detect these changes, especially in remote areas for which there are no good local maps. In future, there may be satellites producing constantly updated topographical maps of the entire globe.

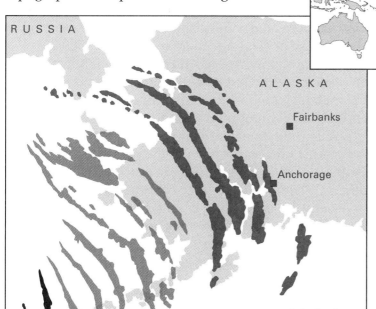

RUSSIA

ALASKA

Fairbanks

Anchorage

Ash cloud
- day 1
- day 2
- day 3

Mount Cleveland

High level hazard *(left)*
Clouds of ash created by erupting volcanoes can be very dangerous to airplanes. The ash can get into jet engines and cause them to shut down. The Alaska Volcanoes Observatory uses satellite data to detect and track ash clouds. It issues up to the minute warnings to aircraft flying over the Bering Strait region *(above)*. This is one of the world's busiest air traffic lanes, and also one of the most volcanically active areas on Earth. The map on the left shows the progress of an ash cloud produced by an eruption of Mt. Cleveland in February 2001.

© DIAGRAM

253

13 | EARTH'S GRAVITY FIELD

Tropical gravity *(below)*
Earth's shape makes a difference to the gravity at different locations. Because Earth is a slightly flattened sphere the poles are closer to the center of the planet's mass than the equator. A person standing on the equator will experience about 0.5 percent less gravity than a person at the North Pole because he is further from Earth's center of mass.

Gravity is not the same everywhere on Earth. At some locations a person weighs about one percent more or less than he or she does at other locations. This is because local gravitational attraction is slightly stronger or weaker in different places. The variation is so small that nobody could ever feel the difference, but it is big enough to be important to scientists studying the Earth.

Local gravitational attraction varies because Earth is not made of the same material throughout its mass. For example, some forms of rock are much denser than others. Where there are large concentrations of unusually dense rock the gravity field is stronger than in places where less dense rock is common.

Geologists sometimes use weights attached to very accurate scales to measure whether an area has particularly high or low density rocks. These methods are used when prospecting for particular minerals. Satellite technology has now allowed the gravity field of the entire globe to be mapped. In 2002 a NASA mission known as the Gravity Recovery and Climate Experiment (GRACE) was launched. GRACE consisted

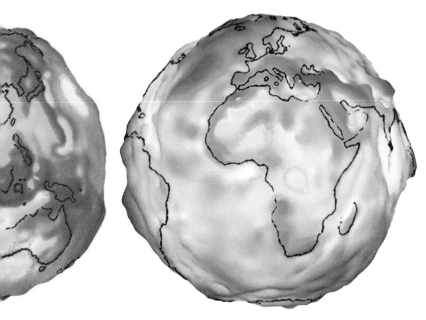

Bumpy Earth *(left)*
These globes exaggerate the uneven nature of Earth's gravity field. "Hills" are areas where gravity is stronger than average and "dips" are where it is weaker than average.

GRACE *(below)*
The Gravity Recovery and Climate Experiment (GRACE) consists of two identical satellites flying about 137 miles (220 km) apart. The satellites constantly measure the distance between each other to detect the tiny changes caused by differences in Earth's gravity as they orbit the planet at an altitude of 311 miles (500 km).

of two identical satellites designed to orbit Earth while flying in close formation. By monitoring the position of the two satellites relative to Earth and to each other constantly, scientists were able to detect the tiny changes in their orbits caused by differences in the gravity field of the ground below.

The results of GRACE are important for two reasons. Firstly, they allow scientists to subtract the effects of gravity field variation when observing other effects. For example, when measuring differences in sea level caused by temperature changes it is important to know where distortions are also being caused by gravity differences. Secondly, these data allow scientists to observe changes in the pattern of the gravity field over time. Continued measurements of the gravity field have shown that the Antarctic region has lost large amounts of mass since 2002. This supports other observations that the Antarctic ice sheet is shrinking.

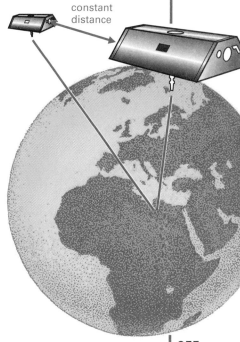

constant distance

13 EARTH'S MAGNETIC FIELD

True north?

This map of Earth's magnetic field in the Northern Hemisphere demonstrates how misleading a compass can be. Known as a declination map it shows how many degrees away from the geographical north pole a magnetic compass will point depending on where the compass is. For example, at point **A** in Alaska a compass will point between 20 and 25 degrees to the east (+20 to +25 degrees declination) of true north. At point **B** in Greenland, a compass will point between 30 and 35 degrees to the west (−30 to −35 degrees declination).

Earth has a magnetic field that reaches thousands of miles into space. The area of influence of this magnetic field is known as the magnetosphere. Earth's magnetic field is thought to be produced by the complex motion of molten conducting material around the planet's core. It can be thought of as a giant bar magnet with one end near the North Pole and the other near the South Pole. In fact, Earth's magnetic field is much more complex than a bar magnet and varies from place to place and over time.

The magnetic north pole is the point on the globe that the north end of a compass needle will point toward. It is near to, but not the same as, the geographic north pole (or "true north"). Both the magnetic north and south poles are constantly in motion. They usually shift position by at least nine miles (15 km) in a year. Unlike a bar magnet, the north and south poles of Earth's magnetic field are not directly opposite each other. Geologists have also discovered that the magnetic north and south poles periodically switch positions. Every 250,000 years, on average, there is a complete reversal of the field. The last reversal took place more than 700,000 years ago.

Scientists monitor Earth's magnetic field for two main reasons. Firstly, local variations in the field can provide information about the geology of an area. Secondly, the field protects all life on Earth from potentially dangerous radiation from the Sun. Interactions between Earth's magnetosphere and solar radiation can result in

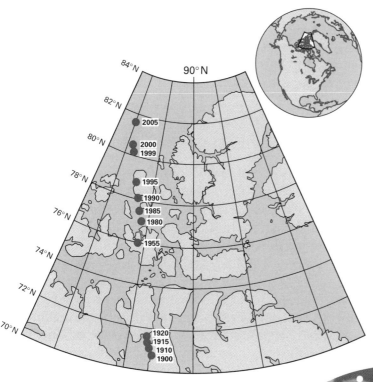

Wandering poles
Over the course of the twentieth century, the magnetic north pole migrated steadily northwest across the islands to the north of Hudson Bay, Canada. It has traveled about 685 miles (1,100 km) in 100 years. At its current rate of migration, the magnetic north pole will be in Siberia, Russia by 2050.

magnetic storms. These events can interfere with satellite sensors, interrupt global communications, and may expose astronauts and pilots of high-altitude aircraft to increased radiation hazards.

Monitoring stations
The United States Geological Survey (USGS), among other institutions, has been monitoring Earth's magnetic field since 1843. Measurements from other parts of the world are available from as far back as 1600. The USGS currently has 14 permanent geomagnetic observatories scattered across the Northern Hemisphere.

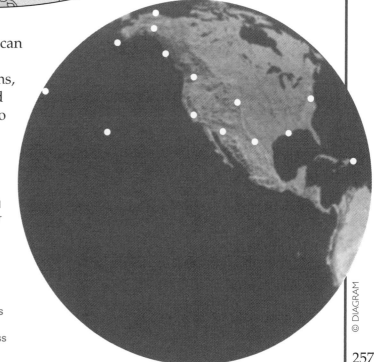

© DIAGRAM

13 | FAULTS AND EARTHQUAKES

Taking the strain *(right)*
In this laser strainmeter, a laser beam is split into two parts at the first station. The first part shines directly into a photodetector while the second part travels several hundred feet (meters) to a mounted mirror on the second station where it is reflected back to the photodetector. The two parts of the beam interfere in a recognizable pattern. If the distance between the two stations changes, the interference pattern also changes.

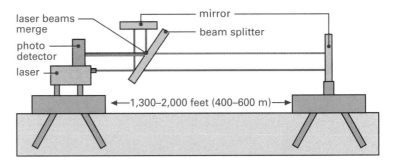

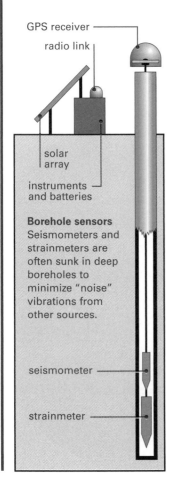

GPS receiver

radio link

solar array

instruments and batteries

Borehole sensors
Seismometers and strainmeters are often sunk in deep boreholes to minimize "noise" vibrations from other sources.

seismometer

strainmeter

EarthScope is a joint U.S. Geological Survey and NASA program to learn more about the evolving geology of the North American continent. It involves three main branches known as the USArray, the Plate Boundary Observatory (PBO), and the San Andreas Fault Observatory at Depth (SAFOD).

USArray is a continent-wide network of seismometers and GPS receivers. When the array is fully operational there will be about 3,000 permanent and mobile seismic stations in the network. About 400 of these will form a regularly spaced grid across the continent. Others will be concentrated around areas of special interest, mostly in the western United States and Alaska.

The PBO is a network of more than 1,000 GPS receivers and strainmeters designed to monitor the movements of the Pacific and North American plates relative to one another. A strainmeter is an instrument that measures an object's changing shape in response to the application of force. Earth's rocks are subject to many forces, especially at plate boundaries. Measuring stretching or compression in rocks at these locations tells scientists how fast plates are moving and in which direction.

SAFOD involves drilling a 1.86 mile (3 km) deep hole into the heart of the geology that makes up the San Andreas Fault in California. Instruments have been positioned deep underground and core samples taken. The SAFOD drill site is located in the region of Parkfield, the epicenter of five strong earthquakes since 1857.

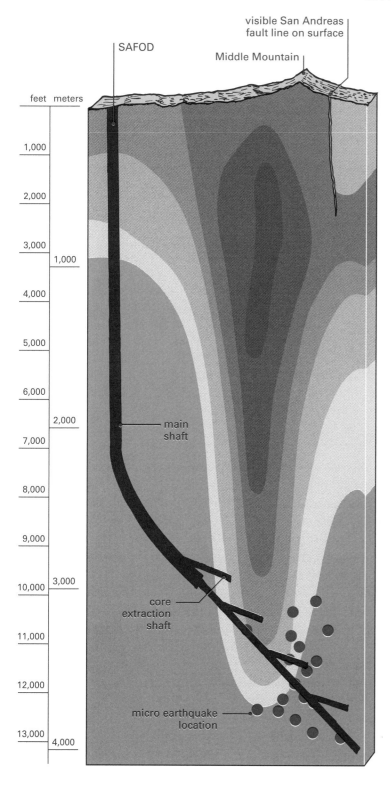

visible San Andreas
fault line on surface

SAFOD

Middle Mountain

main
shaft

core
extraction
shaft

micro earthquake
location

SAFOD hole
SAFOD scientists drilled a
vertical shaft to a depth of
about 1.4 miles (2.2 km).
Advanced drilling techniques
used in oil prospecting were
then used to extend the shaft
downwards and laterally into
the fault zone. At various
points along this shaft,
810-foot (250 m) long core
samples of the rocks were
taken.

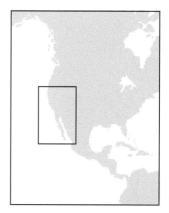

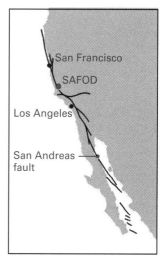

San Francisco

SAFOD

Los Angeles

San Andreas
fault

© DIAGRAM

259

13 | SEISMIC MONITORING

Seismology is the study of earthquakes and the way that the waves they produce travel through Earth. The behavior of these waves provides a useful way of sensing the structure of Earth's interior.

Wave types
P-waves are a back and forth motion that forms alternating areas of compression and dilation. S-waves are an up-and-down motion similar to that of ocean waves.

P-wave

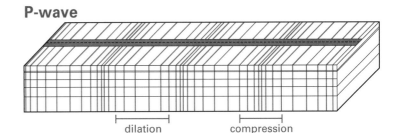

dilation   compression

S-wave

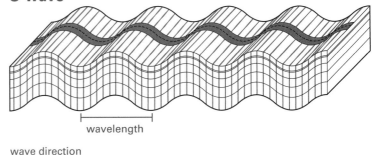

wavelength

wave direction

Locating earthquakes
Earthquakes are very common along the Mid-Atlantic Ridge. This diagram shows how data from three seismic recording stations can be used to pinpoint the epicenter of an earthquake. P-waves travel faster than S-waves and the interval between them increases with distance. The interval between the arrival of the P-waves and the arrival of the S-waves reveals how far away the earthquake is from the station that detects them. Knowing the distance from three stations allows a location to be pinpointed.

Waves spread out from an earthquake in a similar way to the waves that spread out from a pebble dropped into water. These waves travel through rock. They cause the rock to move, just like water waves cause water to move. It is this movement that destroys buildings. Earthquake waves are so powerful that they can be detected by sensitive instruments on the other side of the world from the center of the earthquake. Primary waves (P-waves) and secondary waves (S-waves) are the most useful for studying Earth's interior.

Waves of all kinds are affected by the material they pass through. By sensing the way that waves travel though the planet, seismologists were able to determine that some regions must be solid and others liquid. Scientists now know that Earth has a very dense solid core surrounded by a liquid or semi-liquid outer core encased in a thick, mainly solid mantle.

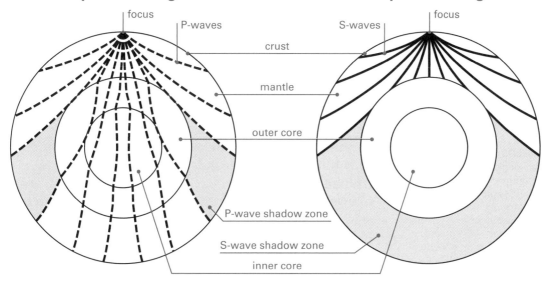

P-waves' paths through Earth

focus
P-waves
crust
mantle
outer core
P-wave shadow zone
S-wave shadow zone
inner core

S-waves' paths through Earth

S-waves
focus

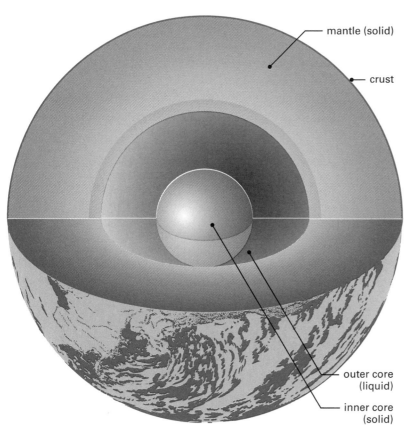

mantle (solid)
crust
outer core (liquid)
inner core (solid)

Wave paths

P-waves will travel through solids and liquids, S-waves do not travel through liquids. Waves are refracted when they pass from one material into another material with a different density, and they also change speed. This also happens to light when it passes from air into glass or water.

The S-wave shadow occurs because S-waves will not pass through Earth's molten (liquid) outer core.

The P-wave shadow occurs because P-waves are refracted when they pass from the solid mantle into the molten outer core, and again when they pass back into the mantle.

© DIAGRAM

SATELLITE LASER RANGING

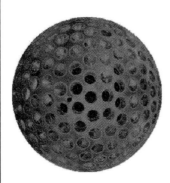

LAGEOS 1
This dedicated satellite laser ranging vehicle is about 24 inches (60 cm) in diameter and has 426 laser reflectors on its outer surface. It weighs 893 pounds (405 kg) and orbits Earth at an altitude of about 3,666 miles (5,900 km) every 225 minutes.

Satellite laser ranging is a very accurate method for measuring the movement of tectonic plates and other changes in the shape of Earth's surface. It is also of crucial importance in finding the precise orbit of an Earth observation satellite once it has been launched into space.

Satellite laser ranging (SLR) works by bouncing a very short burst of laser light off a reflector mounted on a satellite in orbit and recording the time taken for it to return. The travel time of the light can be measured to within a few picoseconds (one picosecond is one trillionth of a second or 0.000000000001 seconds). This allows the distance from the laser ranging station to the satellite to be measured to within about 0.12 inches (0.3 cm). This very high accuracy allows scientists to detect the movement of the continents that fixed laser ranging stations are built on. The International Laser Ranging Service has more than 60 laser ranging stations around the world.

There are several satellites dedicated to SLR. The LAsar GEOdynamics Satellites (LAGEOS) are an example of dedicated SLR craft. *LAGEOS 1* and *2* are small, very

Reflector
SLR satellites are fitted with banks of laser reflectors (also known as retroreflectors). These mirrored prisms always reflect a beam of laser light at exactly the same angle as it arrives. This ensures that the returning beam can be sensed by the laser ground station.
a incoming laser beam
b reflected laser beam
c retroreflector

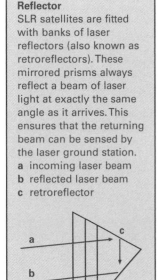

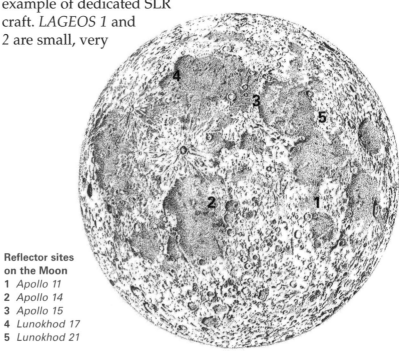

Reflector sites on the Moon
1 *Apollo 11*
2 *Apollo 14*
3 *Apollo 15*
4 *Lunokhod 17*
5 *Lunokhod 21*

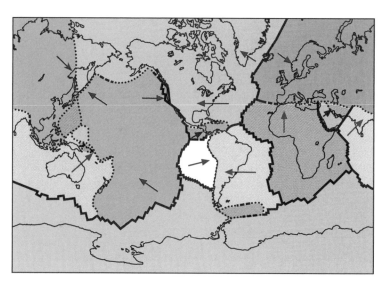

Plate movement
SLR techniques allow planet scientists to measure the very slow movement of tectonic plates. For example, North America and Europe are known to be moving apart at a rate of about 1 inch (2.5 cm) per year. This is due to the spreading ridge through the mid-Atlantic that drives the two tectonic plates apart.

dense spheres covered in laser reflectors. They are very simple craft, having no electrical systems or engines, but are in very stable orbits that should allow them to be useful laser ranging targets for decades. As well as the dedicated satellites, many Earth observation satellites are also fitted with laser reflectors. *TOPEX/Poseidon*, a NASA satellite designed to study the oceans, is a good example. SLR is regularly used to determine the satellite's altitude. This helps to calibrate the instruments on board that rely on having accurate information about altitude in order to measure sea level height.

SLR is also used to monitor the distance between Earth and the Moon. Several laser reflector arrays were placed on the Moon during the American Apollo and the Russian Lunokhod missions.

SLR technique
A fixed laser ranging station (**a**) shoots a pulse of laser light (**b**) so that it intercepts an SLR satellite (**c**) in a fixed orbit (**d**) around Earth. The time taken for the laser light to make the round trip to the satellite and back can be used to calculate the distance between the laser ranging station and the satellite. A year later, the same station repeats the process. The orbit of the SLR satellite has not changed, but the position of the laser ranging station has because it is located on a moving tectonic plate (**e**). The distance between the station and the satellite at the same point on its orbit is less.

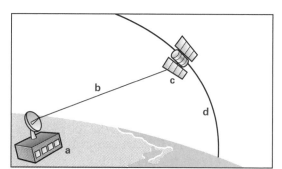

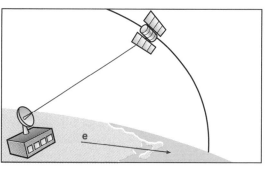

© DIAGRAM

13 PROSPECTING

ASTER's gaze

ASTER monitors 14 different wavelengths of light emitted or reflected from Earth's surface. Every material reflects certain wavelengths and absorbs others. ASTER can "see" what a patch of land is made up of or covered by because it registers which wavelengths are reflected from it and which are not. The chemical composition of rock affects its pattern of reflection. Ores and associated minerals can be identified by their highly specific spectrum of reflectivity. In this image of California's Death Valley ASTER has picked out areas characteristic of quartz-bearing rock (colored areas).

Economic growth depends on a steady supply of minerals. Mining companies provide the majority of these minerals by extracting them from the rocks of Earth's crust. Mining can be a very environmentally damaging process. In the past, vast tracts of land have been devastated by mining operations that have torn away the soil and plants just in order to locate precious ores. Today, developed companies have laws that prevent this wholesale destruction, but the minerals still have to be found. Geologists are now able to use remote sensing technologies that enable them to find Earth's resources without ripping up the fragile layer of life that covers them.

In 1999 NASA launched its *Terra* satellite. *Terra* is the flagship of NASA's Earth Observation System, a fleet of Earth observation satellites that will provide 15 years of continuous data. One of the instruments aboard *Terra* is the Advanced Spaceborne Thermal Emissions and Reflection Radiometer (ASTER). ASTER is now being used by the United States Geological Survey (USGS) office as part of a project to find large new deposits of valuable minerals around the world. The project, known as the Global Mineral Resource Assessment Project, is using the new technology to survey vast areas of Earth that it would have taken aircraft or ground-based geologist decades to examine.

large
medium
small

The Global Mineral Resource Assessment Project is concentrating on locating copper, platinum-group metals, and potash. Copper is the most valuable and widely used of the non-precious metals because it is so important to the electronics industry. Platinum and similar metals are important catalysts in the petrochemical and automobile industries, and there are currently few known sources. Potash is a vital fertilizer for food production.

Mineral ores are rarely visible on the surface. Instead, geologists must look for signs that ores lie below the surface. These may be other minerals that are associated with the sought after metal. For example, the geological processes that create copper ore also tend to produce an abundance of iron sulfide, muscovite, and alunite. Iron sulfide reacts with oxygen in the air to produce the red coloring common in rivers and streams near copper mining areas. Some metal ores can also be spotted because of the effect they have on plant life. Heavy metals such as mercury and lead tend to damage plant life when they are taken up through the roots. Remote sensing can identify these variations in plant cover.

Rich seams *(above)*
The Global Mineral Resource Assessment Project has predicted that major new deposits of valuable minerals are waiting to be discovered in the areas shown on this map. Prospectors need to take into account the damage that may be done to the biodiversity of these areas if mining takes place.

Iron find *(above)*
A satellite image showing iron oxide concentrations (colored areas).

© DIAGRAM

OCEAN SURFACE TOPOGRAPHY

Scanning the surface
Jason 1, like *TOPEX/Poseidon*, uses a radar altimeter to measure ocean surface topography.

The Ocean Topography Experiment is a joint NASA and French Space Agency program to measure the changing topography of the ocean's surface accurately. It has used remote sensing radar instruments mounted on the *TOPEX/Poseidon* satellite and its replacement *Jason1*. *TOPEX/Poseidon* spent 13 years in orbit scanning 90 percent of Earth's ice-free ocean surface every 10–11 days. It produced an unprecedented insight into the workings of global ocean circulation, events such as El Niño, and changes in mean sea level.

The surface of the ocean is not flat. Even if there were no wind or tides to disturb the ocean there would be huge ridges, hills, and valleys distorting its surface. These bumps and dips are spread over such enormous areas that they are invisible to the naked eye, but satellites can record them. They exist because the water of the ocean is not a uniform temperature. Deep in the ocean and around the poles the water is cold. In the tropics and at the surface the water is warmer. Water expands as it warms and fills a greater volume than cold, dense water. A current of cold water welling up from the depths will create a physical valley in the surface of the ocean because it is more compact than the warm water around it.

The surface is also distorted by variation in Earth's gravity field from place to place. These gravity effects must be subtracted from TOPEX/Poseidon's results to see the effects of circulation and mean sea-level change in isolation.

Ocean temperatures *(right)*
August water temperatures vary from below 50°F (10°C) (shaded gray) to above 77°F (25°C) (colored). Areas of water with different temperatures have different densities. As these masses of water move around with the currents, they produce dips and swells in the ocean.

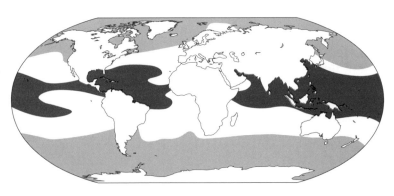

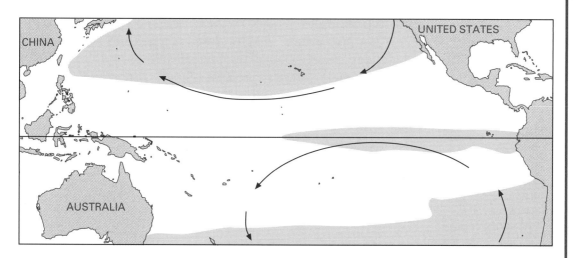

Normal conditions *(above)*
Usually a spike of cold water extends into the Pacific from the coast of South America. This cold water is high in nutrients and supports huge numbers of anchovy.

☐ cold surface water

◄── surface current

El Niño *(below)*
This image of the topography of the Pacific Ocean produced by *TOPEX/Poseidon* in December 1997 clearly shows the influence of the major El Niño event of that year. A ridge of water thousands of miles long and up to 5.5 inches (14 cm) higher than mean sea level stretches out from the coast of South America. This ridge is the current of warm water that periodically disrupts weather all over the world and devastates the Pacific's anchovy fishing grounds.

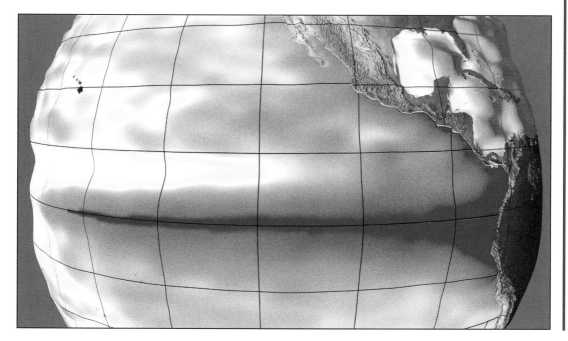

© DIAGRAM

13 | SEAFLOOR PROFILING

The seafloor is a geologically rich and dramatic area of Earth's surface. Vast mountain ranges and incredibly deep chasms lie hidden beneath the oceans. Until the twentieth century, geologists knew almost nothing about the topography and rocks of the ocean floor.

Sonar

Active sonar works by sending out a pulse of sound waves and then detecting the reflections of those waves from underwater objects. Several species of whales and dolphins use a similar technique to sense their environment and find prey.

transmitter/receiver

sound waves

reflected waves (echo)

Sonar (SOund NAvigation and Ranging) was invented during the First World War as a method of detecting submarines. Sonar is similar to radar except that it uses sound rather than microwaves. This is because sound travels as efficiently through water as microwaves do through air. It is a form of active remote sensing. Sonar was first used to detect features of the

Sidescan "fish"

Sidescan sonar is often housed in a torpedo-shaped object known as a fish. The fish is towed behind a ship at a depth that allows for the best resolution images. It is also possible to mount sonar on autonomous underwater vehicles.

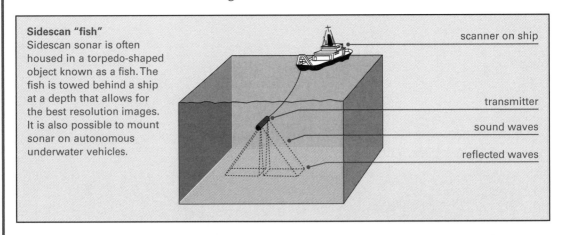

scanner on ship

transmitter

sound waves

reflected waves

seafloor in the 1920s. The first detailed maps of what lay at the bottom of the ocean were produced in the 1950s. It was only then that scientists began to understand how large and geologically significant

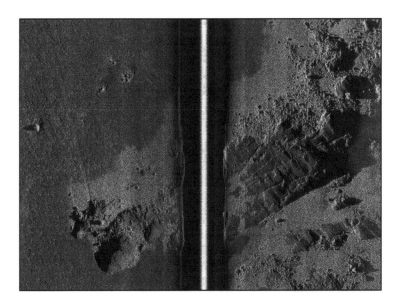

Multibeam sonar image
This high resolution sidescan sonar image of the seafloor clearly shows rock formations protruding from a flat, muddy area. The blank strip down the center of the image is the area directly under the sonar for which no data is collected.

features such as mid-ocean ridges and trenches really were. Data of this kind led to the development of the theory of plate tectonics and a revolution in the understanding of Earth's geology.

As influential as these early seafloor mapping exercises were, they only provided a very rough idea of the seafloor's topography. At the time, sonar measurements could only be taken one at a time and only of the seafloor directly below the ship carrying the equipment. This is like trying to map the Rocky Mountains by surveying only one thin strip of land at a time—it reveals that mountains are there, but tells you nothing about their structure and shape. In the 1960s new technology made it possible to link several sonar instruments together so that a wider area could be surveyed in one pass. This is known as multibeam or sidescan sonar because it uses multiple sonar emitters and because it can "see" sideways from the ship rather than just straight down.

Modern multibeam sonar systems can produce high resolution images of seafloor topography in 1,650-foot (500 m) wide strips, known as swathes. These systems can also provide information about the kind of material an area is made of.

Discovering wrecks
This previously unknown shipwreck was found by a National Oceanic and Atmospheric Administration (NOAA) vessel using multibeam sonar. The wreck is 118 feet (36 m) long and sits in shallow water near Narragansett Bay, Rhode Island. It posed a considerable danger to ocean-going shipping in the area before its position was known.

© DIAGRAM

OCEAN FLOW

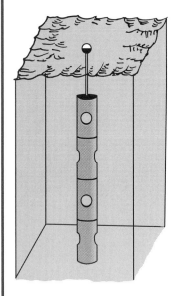

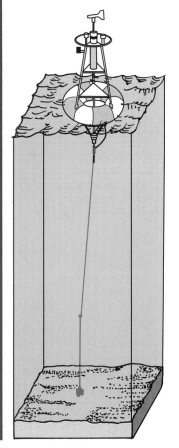

The El Niño/Southern Oscillation (ENSO) phenomenon is a large-scale interaction between the tropical Pacific Ocean and the atmosphere that has a large effect on weather and ocean currents. It involves a slow shift of sea-level atmospheric pressure over a period of two to seven years. When sea-level air pressure is higher than usual over the eastern Pacific, it tends to be lower than usual over the eastern Indian Ocean. This causes strong winds to blow from east to west across the Pacific. When the reverse is true, these winds are much weaker or even blow in the opposite direction. The change in these winds has a strong influence on the ocean surface currents that they drive. El Niño events are occasions when the general flow of the surface currents off the west coast of South America is reversed.

In the last 20 years, a network of sensors has been strung across the Pacific to help planet scientists understand and predict ENSO events. It is known as the ENSO Observing System and consists of moored buoys, tide gauge stations, drifting buoys, and research ships. Data from the network is collected and transferred using the Argos satellite system.

Argos is a satellite system dedicated to the collection and distribution of environmental and Earth science

Global drifters (left top)
Global Lagrangian Drifters are free-floating buoys that are released into the oceans and tracked via GPS or Argos satellites. By releasing many drifters across the oceans scientists are able to track the movement of bodies of water. The buoy consists of two parts. Beneath the surface a suspended "holey sock" drogue ensures that the buoy moves with the currents instead of being pushed by the wind or waves. The floating part contains air pressure, seawater temperature, and seawater conductivity instruments, as well as a transmitter.

Moored buoys (left bottom)
More than 70 Autonomous Temperature Line Acquisition System (ATLAS) buoys are permanently moored in the tropical region of the Pacific Ocean. They make up the Tropical Atmosphere/Ocean (TAO) array, a vital element of the ENSO Observing System. Each ATLAS buoy has instruments to measure weather at the surface as well as seawater temperatures and conductivity at several depths through the water column and pressure. Data is transmitted to the Argos satellite system in real time.

data from a variety of fixed and mobile transmitters. Argos equipment is mounted on polar orbiting satellites operated by the National Oceanic and Atmospheric Administration (NOAA) as well as Japanese and European satellites. These instruments can locate Argos transmitters, such as those attached to drifting buoys, to within a few hundred feet and upload data from them. This data is then downloaded to ground stations and then delivered to scientists all over the globe.

ENSO Observing System *(below)*
The network combines moored buoys, free-floating buoys, tidal-gauge stations and patrolling surface ships that maintain the network and carry out additional measurements.

A Greek god *(above)*
The Ocean *TOPography EXperiment* (*TOPEX*) spacecraft *Poseidon* was named for the ancient Greek god of the oceans. Its successor, *Jason1*, was named for the sea-voyaging hero of the Greek legend, Jason and the Argonauts.

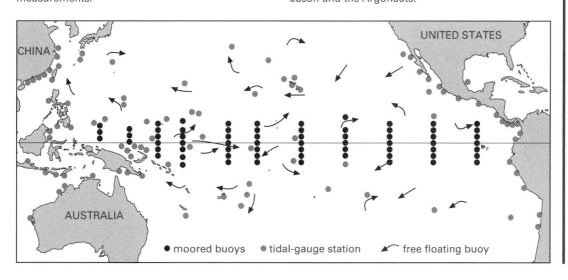

UNITED STATES

CHINA

AUSTRALIA

● moored buoys ● tidal-gauge station ⤺ free floating buoy

© DIAGRAM

13

THE RING OF FIRE

Ring of Fire
Also known as the circum-Pacific Seismic Belt, the Ring of Fire extends for 25,000 miles (40,000 km) from the southern tip of South America up to the Aleutian Island chain and down to the southern tip of New Zealand. It is the world's most active earthquake and volcano zone.

1 Aleutian trench
2 Kurile trench
3 Izu Bonin trench
4 Ryukyu trench
5 Mariana trench
6 Philippine trench
7 Java (Sunda) trench
8 Bougainville trench
9 Tonga trench
10 Karmadec trench
11 Middle America trench
12 Peru-Chile trench

The Ring of Fire is a highly active zone of seismic and volcanic activity that runs around the perimeter of the Pacific Ocean. Ninety percent of the world's earthquakes occur in this zone as well as the majority of the world's volcanic eruptions. It is a nearly continuous series of oceanic trenches, island arcs, and volcanic mountain ranges associated with fault lines where crustal plates are interacting. Most of these active areas lie on the seafloor.

Scientists use a number of methods to study these active zones. Seafloor observatories are one method. The New Millenium Observatory (NeMO) is one such research station located on the Juan de Fuca ridge fault zone off the west coast of the United States. Deep sea submersibles and remotely operated underwater

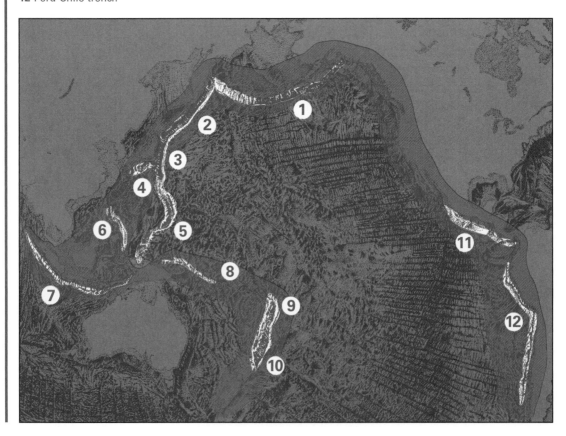

vehicles are also used to explore these areas. A series of expeditions using these vehicles has been undertaken by the National Oceanic and Atmospheric Administration (NOAA) along stretches of the Ring of Fire. Studies of seismic activity across the whole of the ocean are carried out using the US Navy's SOund SUrveillance System (SOSUS)—a network of highly sensitive sonar receivers that are able to detect the sounds made by seafloor eruptions and earthquakes over thousands of miles.

NeMO is a seafloor observatory on the Axial Seamount—an active seafloor volcano about 250 miles (400 km) off the coast of Oregon. The NeMO program is designed to make a detailed study of the geology, chemistry, and biology of a highly active region of a mid-ocean ridge over the course of several years. NeMO consists of permanent seafloor instruments that record seismic activity coupled with annual expeditions made by scientists aboard a surface research vessel. During these expeditions, remotely operated vehicles such as the Canadian *Remotely Operated Platform for Ocean Science* (*ROPOS*) are used to make observations of the area and to take geological and chemical samples.

NeMO location
NeMO sits on one of the most active regions of the Ring of Fire in the eastern Pacific.

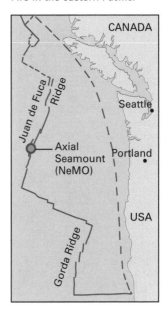

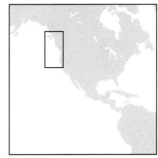

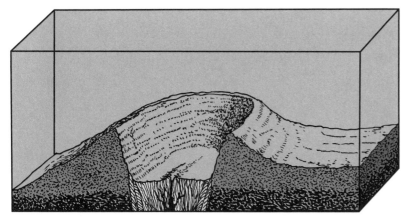

Axial Seamount *(left)*
The Axial Seamount, site of NeMO, is one of the most active seafloor volcanoes in the eastern Pacific. It rises 3,600 feet (1,100 m) above the surrounding seafloor and is at a depth of more than 4,500 feet (1,400 m). It has a distinctive horseshoe shaped caldera that is defined on three sides by a 495-foot (150 m) near-vertical boundary fault.

© DIAGRAM

13 | ACOUSTIC MONITORING

SOFAR Channel

The SOund Fixing And Ranging (SOFAR) channel exists because the ease with which seawater transmits sound varies depending on its temperature and pressure. This has the effect of refracting, or bending, sound waves so that they stay within a narrow band at depths of between about 1,640 feet (500 m) to 3,280 feet (1,000 m). Sound waves can "bounce" along inside this channel for thousands of miles, rather like an echo in a narrow gorge.

Since 1990, the National Oceanic and Atmospheric Administration (NOAA) has been using the Navy's SOund SUrveillance System (SOSUS) to detect seismic activity on the seafloor. SOSUS was originally developed as a method of tracking military submarines in the Atlantic and Pacific oceans. The NOAA program has successfully turned this sophisticated acoustic monitoring network into a tool for Earth science.

"Acoustic monitoring" refers to the detection and recording of sound waves. In the case of SOSUS, these sound waves are generated by seismic events and travel through seawater. Sound underwater is detected by instruments known as hydrophones, which are similar to microphones. SOSUS allows the NOAA to continually monitor low-level volcanic and earthquake

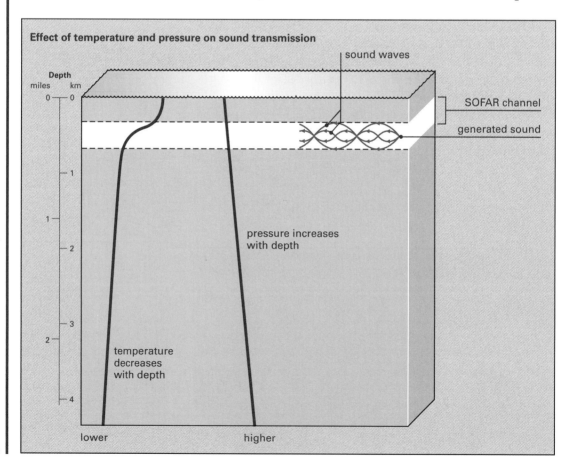

Effect of temperature and pressure on sound transmission

activity across the world's oceans and to sense seismic events in real time remotely. The NOAA has also deployed a network of autonomous hydrophones tethered to the ocean floor. Marine biologists use the same technology to record sounds made by ocean-dwelling creatures such as whales.

Acoustic monitoring in the ocean relies on a phenomenon known as the SOund Fixing And Ranging (SOFAR) channel. SOFAR is a layer in the ocean that tends to transmit sound particularly well. SOSUS instruments and autonomous hydrophones are positioned to take advantage of SOFAR. Some are positioned on seamounts or other features of the seafloor that rise into the SOFAR channel, others are tethered at the most efficient depth.

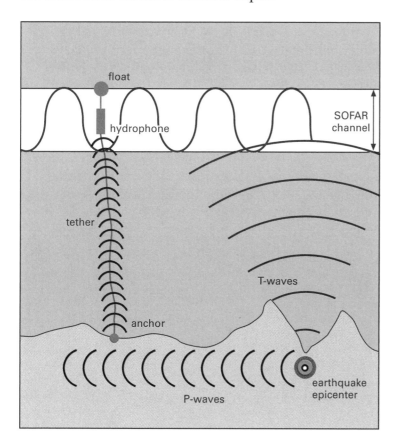

Autonomous hydrophones
A seismic detector suspended in the SOFAR channel detects earthquake activity in two ways. An earthquake on the seafloor creates P-waves that travel through the rocks and along the tether. The same earthquake also creates T-waves—sound waves produced by the release of energy into seawater. These T-waves are transmitted along the SOFAR channel and are picked up by a hydrophone. By measuring the delay between the fast moving P-waves and the slower-moving T-waves, the instruments can calculate how far away and where the earthquake epicenter was. A network of instruments allows the exact location of the earthquake to be pinpointed.

© DIAGRAM

13 SUBMERSIBLES

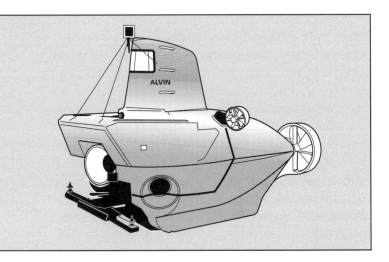

Deep Sea Vehicle (DSV) *Alvin*
Alvin carries a crew of three inside a titanium pressure sphere. It has viewing ports and powerful lights to allow visual exploration of the deep ocean. Mechanical arms and instruments are usually mounted on the front of the submersible to allow the crew to take samples and measurements. *Alvin* can dive to a maximum depth of 14,700 feet (4,500 m).

More than 70 percent of Earth's surface is covered by the ocean. Most of the vast area of the seafloor has never been explored directly. Submarines have been in use for almost 100 years, but even modern military submarines cannot go nearly deep enough to explore most of the seafloor. Since the 1960s several special underwater vehicles, known as submersibles, have been built to investigate the mysteries of the abyss.

The earliest submersibles were manned vehicles. The Navy's Deep Submergence Vehicle (DSV) *Alvin* was one of the first and most successful manned submersibles. Built in 1964, *Alvin* has made thousands of exploratory dives. In 1977, scientists on board *Alvin* were the first

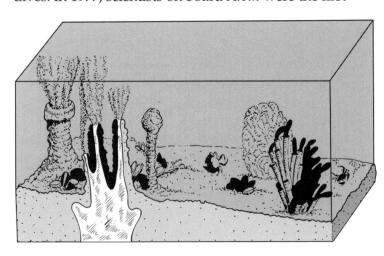

Island of life
A black smoker—a kind of volcanic vent on the deep seafloor. Communities of bacteria and animals around black smokers derive their energy from the heat and minerals expelled by these vents. Unlike other known life forms, they do not rely on the Sun for their energy.

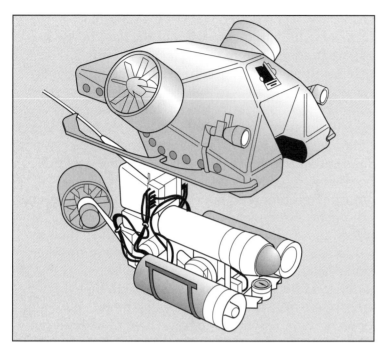

Remotely Operated Vehicle (ROV) *Jason*

There have been two ROVs named *Jason*, both belonging to the Woods Hole Oceanographic Institution. The second *Jason* has been in operation since 2002. It can dive to a maximum depth of 21,400 feet (6,500 m). *Jason* is operated by a three-person crew who "fly" and monitor the vehicle via a cable that runs from a support ship to the ROV.

people to witness and record a kind of volcanic vent on the seafloor known as a black smoker. These vents were found to support communities of plants and creatures that were completely unknown to science. In the 1980s *Alvin* was used to explore the wreck of the *Titanic*. Another manned submersible, the *Trieste*, made the first dive to the deepest part of the seafloor in 1960. It reached a record depth of 36,000 feet (11,000 m) in the Mariana Trench.

With the development of computers and robot technology a new type of unmanned submersible was developed. Some of these submersibles are remotely controlled via a link with an operator aboard a support ship and are known as Remotely Operated Vehicles (ROVs). Others can be programmed to explore an area and then released to carry out their missions under their own control. These are known as Autonomous Underwater Vehicles (AUVs). ROVs and AUVs are smaller and more maneuverable than manned submersibles. They are also able to stay submerged for longer than vessels that have to support a human crew.

Autonomous Underwater Vehicle (AUV) *Abe*

The AUV *Autonomous Benthic Explorer* (*Abe*) is a fully independent robotic deep-sea explorer. It can be programmed to explore a large area of the seafloor and to take photographs or samples at specific locations or at specific times. The advantage of an AUV is that it does not need an expensive support ship and a human crew to carry out surveys of large areas. *Abe* can operate at depths of 16,400 feet (5,000 m).

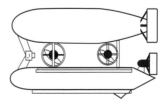

© DIAGRAM

13 MONITORING THE ATMOSPHERE

Earth's atmosphere is crucial to life on Earth and interacts with both the oceans and the land in ways that are fundamental to the behavior of Earth's systems. In recent decades there has been increasing concern that human activities are altering the composition of the atmosphere. These changes are thought to be affecting climate and sea levels globally.

Aura is one of NASA's most advanced Earth Observation System satellites. It is designed to study the chemistry, composition, and dynamics of the atmosphere. *Aura* is equipped with a range of scanning radiometry instruments that sense the atmosphere in visible, infrared, and microwave bandwidths. It also has an instrument dedicated to measuring atmospheric ozone. Data from *Aura* will help scientists to understand the changes that take place in the atmosphere over time and to develop more sophisticated prediction models. These models allow us to see precisely what the long-term effects of climate change will be. *Aura* is one of the latest in a long line of atmospheric observation satellites.

All the energy that drives the circulation of the atmosphere and the oceans comes from the Sun. Planet scientists need to measure the amount of solar radiation that strikes Earth's atmosphere in order to understand these processes. Only instruments on satellites orbiting just outside the atmosphere are able to carry out these measurements. *Nimbus 7*, launched in 1978, was the first satellite to do so. This and other missions have revealed that the Sun delivers 1,368 watts per square meter (W/m²) to the top layer of the atmosphere, though this is

Water vapor
This satellite image of water vapor in the atmosphere is a good example of the kind of data collected by atmospheric observation satellites. Water vapor is the most important natural component of the greenhouse effect.

not constant. The Sun has a well-documented eleven-year cycle during which its output rises and falls, but there are other variations in cycles that are not fully understood. These cycles have a strong influence on Earth's weather systems.

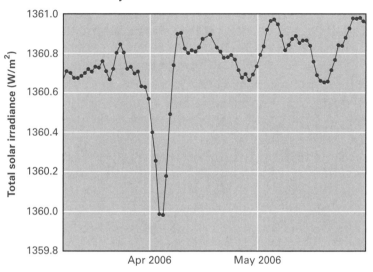

Solar variation *(left)*
Radiation from the Sun arriving at Earth's atmosphere varies from day to day. This graph shows variation in solar radiation over a typical three-month period.

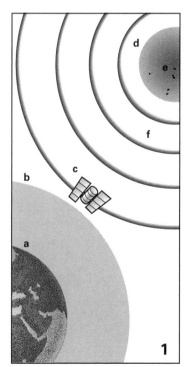

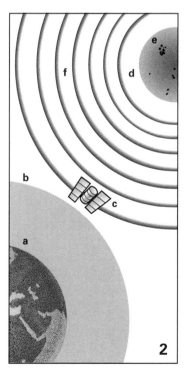

SORCE *(left)*
The *SOlar Radiation and Climate Experiment* (*SORCE*) satellite looks toward the Sun rather than Earth. It records the amount of solar radiation arriving at Earth's atmosphere.
1 During periods with few sunspots, solar radiation is near its minimum.
2 During periods with many sunspots, solar radiation increases. This is because the areas around dark sunspots are actually much brighter than the surface of the Sun usually is.
a Earth's surface
b top of Earth's atmosphere
c *SORCE* satellite
d Sun
e sunspot
f solar radiation

OZONE MAPPING

Ozone plays a vital role in preventing harmful solar radiation from reaching Earth's surface. In the last quarter of the twentieth century, scientists became concerned that the amount of ozone in the atmosphere may be decreasing. The advent of remote sensing satellites has allowed planet scientists to measure the quantity of atmospheric ozone and to determine that it is, in fact, diminishing.

Ozone is a naturally occurring but rare form of oxygen. The ozone layer is a layer of Earth's atmosphere within which ozone concentrations are at their highest. It is found in the stratosphere at altitudes of between 10 and 25 miles (16–40 km). Ozone is formed when ultraviolet radiation from the Sun breaks up normal oxygen molecules in the atmosphere. Some of these recombine to form ozone molecules, which are also broken up by ultraviolet radiation. This process absorbs about 90 percent of the ultraviolet radiation that strikes Earth's atmosphere, preventing it from reaching the ground. This is important because ultraviolet radiation can cause damage to living things. It is for this reason that people wear sunblock in sunny weather.

Falling ozone
Levels of ozone in the stratosphere have been falling ever since continuous monitoring began in 1979. Ozone concentrations are measured in Dobson units (DU). 1 DU of ozone would form a layer 0.0004 inches (10 µm) thick around the whole Earth at standard atmospheric temperature and pressure.

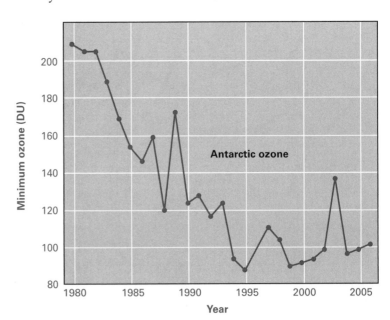

Human-made chemicals introduced into the atmosphere by industrial processes can interrupt the formation of ozone in the stratosphere. Some of these chemicals are known as chlorofluorocarbons (CFCs). Most of them have now been banned by international agreements.

In 1978, NASA launched the *Nimbus 7* satellite. It was equipped with an instrument known as the Total Ozone Mapping Spectrometer (TOMS). TOMS was able to provide a worldwide map of the total amount of ozone in the atmosphere that was updated every couple of days. *Nimbus 7* provided ozone mapping for 14 years until 1993. More TOMS instruments were later flown on Russian, Japanese, and other NASA satellites to maintain a continuous watch on global ozone levels.

Ozone hole
The "ozone hole" over Antarctica is an area of severely reduced ozone levels that appears over the continent each spring. By the 1980s it was dramatically increasing in size.

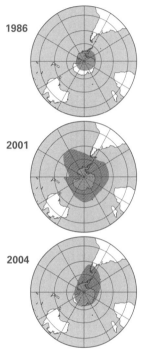

1986

2001

2004

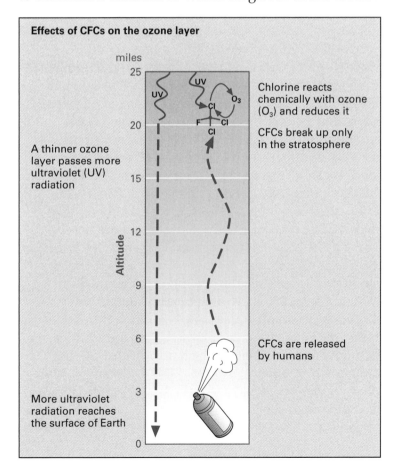

Effects of CFCs on the ozone layer

Chlorine reacts chemically with ozone (O_3) and reduces it

CFCs break up only in the stratosphere

A thinner ozone layer passes more ultraviolet (UV) radiation

More ultraviolet radiation reaches the surface of Earth

CFCs are released by humans

Ozone–oxygen cycle (left)
The constant destruction and reforming of O_3 and O_2 molecules driven by ultraviolet radiation is known as the ozone–oxygen cycle. This cycle uses up the energy of ultraviolet radiation arriving in the atmosphere, thereby preventing it from reaching Earth's surface. Chlorine released by the breakup of CFCs can disrupt the ozone–oxygen cycle, allowing more ultraviolet radiation to pass through.

© DIAGRAM

13 FUTURE DEVELOPMENTS

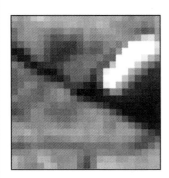

Spatial resolution
Individual image pixels are visible in this highly magnified view of a satellite image.

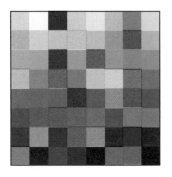

Spectral resolution
Radiometric instruments sense a certain number of bandwidths or "colors."

Radiometric resolution
A fixed number of intensities or "shades" can be sensed for each bandwidth.

Remote sensing of Earth provides data that is immensely valuable to theoretical scientists and to those involved in practical applications of science such as mineral prospecting and weather forecasting. Remote sensing has revealed the true extent of the changes that are currently taking place in Earth's atmosphere, and the effects of these changes on ice cover and sea levels. It is only by thoroughly understanding Earth's interconnected systems that human society can hope to minimize or reverse any damage that has already been done. Remote sensing is by far our most effective tool in that vital enterprise.

There are three key elements that determine the quality of remote sensing data—spatial resolution, spectral resolution, and radiometric resolution. Spatial resolution refers to the size of the individual pixels that make up a remotely sensed scan. Spatial resolution is comparable to the pixel resolution of a digital photograph. Spectral resolution refers to the number of wavelength bands that are sensed by an instrument. Spectral resolution can be thought of as the number of colors that the instrument can "see." Radiometric resolution refers to the number of different intensities of radiation that can be sensed in each wavelength band. It can be thought of as the number of different shades of each color that can be sensed. Earth observation satellites of the future are likely to have improved characteristics in all of these areas.

In 2006 there were more than 70 operational Earth observation satellites carrying instruments to study everything from volcanic eruptions to ozone. From 2006 through 2020 there are plans to launch 100 additional Earth observers. An increasing number of these will be operated by private companies rather than government agencies such as NASA. Building and launching satellites of this kind is already a multi-billion dollar business. The data they produce is becoming increasingly valuable.

Some planned Earth observation missions are:

2008 *Ocean Surface Topography Mission* (*OSTM*)
The OSTM will provide a replacement for the highly successful *TOPEX/Poseidon* and *Jason 1* missions. It will carry instruments to measure ocean surface height.

2008 *Orbiting Carbon Observatory* (*OCO*)
Designed to measure the amount and distribution of carbon dioxide in Earth's atmosphere, OCO will be the first satellite of its kind.

2008 *Atmospheric Dynamics Mission* (*ADM*) *Aeolus*
ADM Aeolus will provide unprecedented data about the strength, direction, and distribution of winds in Earth's atmosphere.

2009 *Aquarius*
Aquarius will carry out global measurements of the amount of salt in the top layer of the ocean (sea surface salinity). This will provide information about the amounts of freshwater flowing into the ocean and levels of evaporation.

2010 *Deep Space Climate Observatory* (*DSCOVR*)
DSCOVR will be positioned at Lagrange 1 (the point of gravitational balance between Earth and the Sun) from where it will have a constant view of the entire sunlit side of Earth.

2013 *Global Precipitation Measurement* (*GPM*)
GPM will continue the work of the *Tropical Rainfall Measuring Mission* launched in 1997 by extending measurements of precipitation from the tropical and subtropical regions to the entire globe.

CHAPTER 14 WHO AND WHERE?

This section provides a wealth of information for readers who wish to find out more about the Earth sciences. It includes a brief survey of pioneers who helped us read the rocks, a guide to museums and collections where stunning examples of rocks and minerals can be seen close up, and a list of Web sites that provide today's gateways to the world of geology.

Illustrations from the works of pioneers in geology
1 Eduard Suess: Folded menilite shales, Moravia. (From *The Face of the Earth*, 1904)
2 Sir Roderick Impey Murchison: Studying rocks in the Urals, Russia. (From *Siluria*, 1867)
3 Sir Charles Lyell: Boulders transported by ice on the shores of the St Lawrence, Canada. (From *Principles of Geology*, 1853)

2

GREAT GEOLOGISTS 1

These pages list geologists and others who helped reveal Earth's shape and structure, or who worked out how its surface features formed. No list this short can be complete. Ours takes in mere dozens of the hundreds of North American, European, and other rock detectives who have pieced together the puzzle of our planet's past.

Georgius Agricola and an illustration from his book on mining *De re metallica*

Alexandre Brongniart

Leopold von Buch

Agricola, Georgius (1494–1555) German father of mineralogy. He studied ores and recognized that mineral veins were deposits left by rising solutions.

Airy, Sir George Biddell (1801–92) British astronomer who, in the 1850s, laid a basis for the theory of isostasy.

Alberti, Friedrich August von (1795–1878) German geologist who named the Triassic System in 1824, from a tripartite division of rocks.

Arduino, Giovanni (1714–95) Father of Italian geology who coined the name "Tertiary," later given to a rock system and geological period.

Barrell, Joseph (1869–1919) American geologist who declared that much sedimentary rock did not form under oceans, and molten magma interacting with the crust formed fissures and new surface features. He coined the terms "lithosphere" and "asthenosphere" and was the first geologist to realize the potential of radioactive dating fully.

Barrow, George (1853–1932) Scottish geologist who found in the 1880s that different temperatures produced different metamorphic rocks from the same ingredients.

Benioff, Hugo (1899–1968) American seismologist whose studies revealed the Benioff zone where earthquake foci descending from an ocean trench into the mantle mark subduction of an oceanic plate.

Bertrand, Marcel-Alexandre (1847–1907) French geologist who showed that the formation of mountain ranges such as the Alps involved massive folding of Earth's crust.

Beyrich, Heinrich Ernst (1815–96) German paleontologist who, in 1854, introduced the term Oligocene.

Bolin, Bert R. J. (born 1925) Swedish pioneer of climate change studies who helped focus world attention on the potential hazards posed by greenhouse gases and acid rain.

Bowen, Norman Levi (1887–1956) American experimental petrologist who showed the order in which minerals crystallize from basaltic magmas as these cool, so producing different igneous rocks.

Broecker, Wallace S. (born 1931) American geologist who pioneered the use of geology to understand the ocean's role in climate change.

Brongniart, Alexandre (1770–1847) French geologist who, in 1829, coined the name "Jurassic ground" for rocks later put in the Jurassic System.

Bryan, Kirk (1888–1950) American geologist and geomorphologist who studied arid-climate landform evolution, and with South African geomorphologist Lester King pioneered the theory of parallel slope retreat often credited to the Austrian Walther Penck.

Buch, Leopold von (1774–1853) German geologist who produced a geological map of Germany in 1824. He studied volcanoes and coined the term "andesite" for Andean volcanic rock of the type now known to be produced at subduction zones.

Carey, Samuel Warren (1911–2002) Australian geologist who proposed the controversial theory that Earth is expanding.

Conybeare, William Daniel (1787–1857) British geologist who with William Phillips in 1822 called certain strata "Carboniferous."

Crutzen, Paul J. (born 1933) German authority on atmospheric chemistry who made landmark discoveries on stratospheric ozone and humanity's capacity to alter Earth's atmosphere.

Cushman, Joseph (1881–1949) American paleontologist who began using foraminiferans for relative rock dating.

Cuvier, Georges (1769–1832) French anatomist and paleontologist whose discovery that many fossil invertebrates were now extinct aided the correlation and relative dating of sedimentary rocks.

Dana, James Dwight (1813–95) American scientist who classified minerals, coined the term "geosyncline," studied coral-rock formation, and theorized about the evolution of Earth's crust.

Dansgaard, Willi (born 1922) Danish paleoclimatologist who demonstrated that measurements of trace isotopes in accumulated glacier ice could be used as an indicator of climate change over time.

Darwin, Charles Robert (1809–82) British naturalist who developed the theory of biological evolution by natural selection. He rightly judged that seabed subsidence created coral atolls, and thought repeated earthquakes could build mountain ranges.

Davis, William Morris (1850–1934) American geographer and geologist; a founder of geomorphology (scientific landform studies) who stressed the so-called cycle of erosion.

Deshayes, Gerard Paul (1797–1875) French conchologist whose work on fossil shells helped subdivide the Tertiary Period into epochs.

Desnoyers, Jules (1800–1887) French geologist who in 1829 separated Quaternary from Tertiary rocks.

Dewey, John F. British scientist who with the American John M. Bird in 1970 showed how plate tectonics had created mountain belts.

Dietz, Robert Sinclair (1914–1995) American oceanographer who helped pioneer and name the theory of seafloor spreading.

Du Toit, James Alexander Logie (1878–1948) South African geologist who suggested that Pangaea had split into (northern) Laurasia and (southern) Gondwana, separated by the Tethys seaway.

Dutton, Clarence Edward (1841–1912) American geologist, seismologist, and volcanologist who advanced and named the theory of isostasy.

Ewing, William Maurice (1906–74) American geologist who measured the thickness of Earth's crust under the oceans and showed it to average only four miles (6.4 km) in contrast to the 22–25 mile (35–40 km) thickness of the crust under the continents.

Geer, Gerard Jakob, Baron de (1858–1943) Swedish geologist who pioneered varve-counting to date sediments in glacial lakes.

Gilbert, Grove Karl (1843–1918) An American founder of landform

Samuel Warren Carey

Clarence Edward Dutton

William Maurice Ewing

Grove Karl Gilbert

© DIAGRAM

14 GREAT GEOLOGISTS 2

Harry Hammond Hess

James Hutton

Sir Charles Lyell

Sir Roderick Impey Murchison

studies (geomorphology) who coined the term "orogeny" for mountain building.

Gressley, Amanz (1814–65) Swiss geologist who coined the term "facies" for lateral variations seen in rocks of the same age.

Guettard, Jean Etienne (1715–86) French mineralogist who produced arguably the first geologic maps.

Hess, Harry Hammond (1906–69) American geologist and geophysicist who in the 1940s discovered guyots (flat-topped submarine peaks) and in 1960 published arguments for seafloor spreading.

Holmes, Arthur (1890–1965) British geologist and geophysicist who put dates to the geological time scale as early as 1913. In 1928 he argued that convection currents in the mantle moved continental crustal blocks which collided, forming mountains and leaving crustal gaps plugged by new rock produced in ocean basins.

Hutton, James (1726–97) Scottish geologist who pioneered uniformitarianism—belief that forces still at work had caused geological change over a vast span of time. He contributed to the understanding of how igneous rocks are formed.

Lapworth, Charles (1842–1920) British geologist who in 1873 identified the Ordovician System by means of fossil graptolites.

Le Pichon, Xavier (born 1937) French marine geophysicist who with W.J. Morgan in 1968 formulated the plate tectonics hypothesis, involving rigid plates that moved.

Lyell, Sir Charles (1797–1875) British geologist who in the 1830s introduced the terms Eocene, Miocene, Pliocene, Pleistocene, and Holocene (Recent). Modern geology owes much to his *Principles of Geology*.

McKenzie, Dan (born 1942) British geophysicist who with R. L. Parker in 1967 related seafloor spreading, transform faults, and island arcs to (lithospheric) plates with interacting boundaries.

Mohorovičić, Andrija (1857–1936) Croatian geophysicist whose earthquake studies led to the discovery of the Mohorovičić Discontinuity, a boundary between the crust and mantle.

Morgan, William Jason (born 1935) American geophysicist, collaborated with Le Pichon, argued in 1968 that "rigid blocks" (lithospheric plates) slid over a weak asthenosphere that acted as a lubricant. In 1971 he proposed that the Hawaiian Islands grew from a fixed "mantle plume" or hotspot underlying moving plates.

Murchison, Sir Roderick Impey (1792–1871) British geologist who recognized the Silurian and Permian systems.

Oeschger, Hans (1927–98) Swiss climate and environmental scientist whose methods for extracting data from sequential layers of polar ice have enabled scientists to access the geochemical data present in the ice archive.

Omalius d'Halloy, Jean-Baptiste-Julien (1783–1875) Belgian geologist who produced systematic subdivisions of geological formations and gave Cretaceous rocks that name.

Patterson, Clair C. (1922–95) American geochemist who determined the age of Earth.

Penck, Albrecht (1858–1945) German geographer and geologist who founded Pleistocene stratigraphy. He also helped pioneer and reputedly named geomorphology (landform studies).

Penck, Walther (1888–1923) Austrian geomorphologist who argued that straight, convex, and concave hill slopes reflected variations in the balance between rates of land uplift and denudation.

Phillips, John (1800–74) British geologist who in 1840 gave names to the Mesozoic and "Kainozoic" (Cenozoic) eras.

Schimper, Wilhelm Philipp (1808–88) German paleontologist who in 1874 gave the Paleocene Epoch its name.

Schlotheim, Ernst von (1764–1832) German paleontologist, a pioneer in using fossils to find the relative ages of rock layers.

Sedgwick, Adam (1785–1873) British geologist who named the Paleozoic Era and the Cambrian and (with Murchison) Devonian systems.

Smith, William (1769–1839) So-called father of English geology. He used fossils to identify sedimentary rock layers, and produced the first geological map of England and Wales (1815).

Steno, Nicolaus (1638–86) Danish geologist who grasped that many rocks derived from sediments laid down in sequence in horizontal layers, and contained remains of ancient living organisms. He realized that running water was the main agent shaping landscapes.

Suess, Eduard (1831–1914) Austrian geologist who coined the terms "sima" and "sal" (later changed to "sial") for mantle and crust; located shields and orogenic belts around the world; and argued that northern and southern prehistoric supercontinents ("Atlantis" and "Gondwanaland") had existed, separated by a "Sea of Tethys."

Tungsheng Liu (born 1917) Chinese geologist who pioneered the study of terrestrial sediments as a means of understanding climate change.

Vine, Fred (born 1939) British geophysicist who argued that bands of normal and reverse magnetism in oceanic rocks showed seafloor formed at different times.

Wegener, Alfred Lothar (1880–1930) German meteorologist and geophysicist who in 1912 announced his theory later known as continental drift, involving splitting of a prehistoric supercontinent, Pangaea, into modern continents.

Williams, Henry Shaler (1847–1918) American paleontologist who in 1891 introduced the term "Pennsylvanian" for Upper Carboniferous rocks in North America.

Wilson, John Tuzo (1908–93) Canadian geophysicist who contributed to plate tectonics theory, and in the 1960s coined the geological terms "plates" and "transform faults."

Winchell, Alexander (1824–91) American geologist who in 1870 introduced the term "Mississippian" for Lower Carboniferous rocks in the Mississippi Valley.

William Smith

Eduard Suess

Alfred Lothar Wegener

John Tuzo Wilson

© DIAGRAM

14 MUSEUMS, COLLECTIONS, SOCIETIES 1

Glacier at Mt. Hayes, Alaska

North American Museums

Countless museums, mines, and quarries show collections of rocks and minerals. Listed on these four pages are some of the most significant displays in North America along with their Web sites. The museums and exhibits are listed in alphabetical order by state or province.

Canada

Alberta The Provincial Museum of Alberta
http://www.abheritage.ca/abnature/geological/
   geological.htm

British Columbia M. Y. Williams Planet Earth Museum
University of British Columbia
http://www.eos.ubc.ca/public/outreach/museum.htm

Ontario Royal Ontario Museum
http://www.rom.on.ca/

Saskatchewan Museum of Natural Sciences University of Saskatchewan
http://www.usask.ca/geology/museum.html

United States

Alaska, Nome Carrie McLain Museum
http://www.nomealaska.org/museum

Arizona, Phoenix Arizona Mineral and Mining Museum
http://www.azminfun.com

Arizona, Tucson The University of Arizona Mineral Museum
http://www.geo.arizona.edu/minmus

California, Los Angeles Natural History Museum
http://www.nhm.org/research/minsci/exhibits.htm

California, Sacramento Discovery Museum
http://www.thediscovery.org

Colorado, Boulder University of Colorado Museum
http://cumuseum.colorado.edu/Research/Paleo/

Colorado, Central City Argo Gold Mill and Museum
http://www.historicargotours.com/history.html

Colorado, Colorado Springs Western Museum of Mining
http://www.wmmi.org

Colorado, Denver Denver Museum of Nature and Science
http://www.dmns.org

Eroded rocks at the Grand Canyon, Arizona

Connecticut, New Haven Yale Peabody Museum
http://www.yale.edu/peabody

Florida, Miami Miami Museum of Science
http://www.miamisci.org/ecolinks/geosphere.html

Massachusetts, Springfield Springfield Science Museum
http://www.springfieldmuseums.org/museums/
science

Michigan, Caspian Iron County Museum
http://www.ironcountymuseum.com

Michigan, Houghton The Seaman Mineral Museum
http://www.museum.mtu.edu/

Minnesota, St Paul The Science Museum of Minnesota
http://www.smm.org

Montana, Butte World Museum of Mining
http://www.miningmuseum.org/

Nebraska, Lincoln University of Nebraska State Museum
http://www-museum.unl.edu/

Nevada, Carson City Nevada State Museum
http://dmla.clan.lib.nv.us/docs/museums/tour/mine.htm

New Mexico New Mexico Mining Museum
http://www.grants.org/mining/mining.htm

New Mexico, Portales Miles Mineral Museum
http://www.enmu.edu/academics/excellence/
museums/miles-mineral/index.shtml

New York, New York City
American Museum of Natural History
http://www.amnh.org

New York, New York City Hayden Planetarium
http://www.haydenplanetarium.org

New York, Rochester Rochester Museum and Science Center
http://www.rmsc.org

North Carolina, Ashville Colburn Earth Science Museum
http://www.colburnmuseum.org/

Oklahoma, Norman
Sam Noble Oklahoma Museum of Natural History
http://www.snomnh.ou.edu

Complex channels in the
Mississippi Delta, Louisiana

A meteorite, Hayden
Planetarium, New York City

© DIAGRAM

MUSEUMS, COLLECTIONS, SOCIETIES 2

The hard ancient granite of Mt. Rushmore, South Dakota

Natural sandstone arch at Arches National Park, Utah

Erosion patterns at Bryce Canyon, Utah

Devils Tower, Wyoming is a resistant mass of igneous rock

Oregon, Portland Oregon Museum of Science and Industry
http://www.omsi.edu

Pennsylvania, Philadelphia Wagner Free Institute of Science
http://www.wagnerfreeinstitute.org/museum.shtml

Pennsylvania, Pittsburgh Carnegie Museum of Natural History
http://www.carnegiemnh.org

Rhode Island, Providence Roger Williams Park Museum of Natural History
http://www.osfn.org/museum

South Dakota, Rapid City Museum of Geology
http://museum.sdsmt.edu

Utah, Eureka Tintic Mining Museum
http://www.juabtravel.com/euraka.htm

Utah, Salt Lake City Hansen Planetarium
http://www.hansenplanetarium.net

Virginia, Blacksburg Virginia Museum of Natural History
http://www.vmnh.net

Virginia, Richmond Science Museum of Virginia
http://www.smv.org

Washington D.C. Smithsonian National Museum of Natural History
http://www.mnh.si.edu

Washington, Seattle Pacific Science Center
http://www.pacsci.org

Wyoming, Laramie University of Wyoming Geological Museum
http://www.uwyo.edu/geomuseum/

National and international societies

The Web sites of these organizations provide access to thousands of local mineralogical clubs and associations as well as hundreds of great and small museums and rock collections throughout the world.

American Federation of Mineralogical Societies

An umbrella organization for more than 250 local mineralogical societies affiliated to the following regional bodies:

California Federation of Mineralogical Societies
Eastern Federation of Mineralogical Societies
Midwest Federation of Mineralogical Societies
Northwest Federation of Mineralogical Societies
Rocky Mountain Federation of Mineralogical Societies
South Central Federation of Mineralogical Societies
Southeast Federation of Mineralogical Societies
http://www.amfed.org

The Australian Mineral Collector

A comprehensive collection of links to Australia's regional mineralogical societies as well as all other mineralogical issues on the continent.
http://www.mineral.org.au

European Mineralogical Union

A Web site maintained by the University of Vienna that includes links to major mineralogical societies in Albania, Austria, Belgium, Bulgaria, Croatia, Czech Republic, Finland, France, Germany, Greece, Hungary, Italy, The Netherlands, Poland, Romania, Russia, Slovakia, Spain, Sweden, Switzerland, and the United Kingdom.
http://www.univie.ac.at/Mineralogie/EMU/

International Mineralogical Association

An organization with the stated aim of promoting international cooperation between mineralogical societies.
http://www.ima-mineralogy.org/

The Mineralogical Society of Great Britain and Ireland

Founded in 1876 The Mineralogical Society has an unparalleled range of information and references extending to every quarter of the globe.
http://www.minersoc.org

River erosion in Yellowstone National Park, Wyoming

Chalk cliffs, Dover, England

Giants Causeway, Antrim, Northern Ireland

© DIAGRAM

14 | WEB SITES 1

A huge amount of useful information can be found on the internet. Information on a particular topic may be available through a search engine such as Google (http://www.google.com). Below is a selection of Web sites related to the material covered by this book.

The publisher takes no responsibility for the information contained on these Web sites.

**Berkeley
Geochronology Center**

Berkeley Geochronology Center
A research institution concentrating on Earth's history and the solar system.
http://www.bgc.org

British Geological Survey: Education
Educational resources to popularize geology.
http://www.bgs.ac.uk/education/

BUBL: Earth Science
An index of bibliographical resources for Earth science from the UK's Bulletin Board for Libraries.
http://bubl.ac.uk/link/e/earthsciencelinks.htm

Chemistry: Web Elements Periodic Table
An authoritative reference resource updated as discoveries in the world of chemistry are made.
http://www.webelements.com

Columbia University, Earth Institute: Center for International Earth Science Information Network (CIESIN)
Working at the intersection of the social, natural, and information sciences, CIESIN provides online data and information management essential to Earth science researchers.
http://www.ciesin.org

Common Ground Database
A database of rock and mineral samples from around the world.
http://commonground.mines.edu

Geological Survey of Canada
The Canadian national information and research center for Earth science.
http://gsc.nrcan.gc.ca

Geology.com
Useful resources for geologists, some commercial.
http://geology.com

Google Earth
A remarkable resource that attempts to provide high-resolution digital mapping of the entire globe.
http://earth.google.com

International Seismological Centre
A nongovernmental organization concerned with the collection, analysis, and publication of standard earthquake information from all around the world.
http://www.isc.ac.uk

Mission to Geospace
An educational resource from NASA that locates the study of Earth in the wider context of the solar system.
http://www-istp.gsfc.nasa.gov/istp/outreach/

NASA: Destination Earth
NASA's stated goal in Earth science is to observe, understand, and model the Earth system to discover how it is changing, to better predict change, and to understand the consequences for life on earth.
http://www.earth.nasa.gov

National Geophysical Data Center: Natural Hazards
Online resources and information about natural hazards and disasters, including searchable databases.
http://www.ngdc.noaa.gov/seg/hazard/

14 WEB SITES 2

National Oceanic and Atmospheric Administration (NOAA)
NOAA conducts research and gathers data about the global oceans, atmosphere, space, and the Sun for practical use by governmental agencies.
http://www.noaa.gov

Natural Disaster Reference Database
A bibliographic research resource for the use of satellites in the analysis and mitigation of natural disasters.
http://ndrd.gsfc.nasa.gov

Natural Resources Conservation Service: Soils
Scientifically-based information about soils from the Department of Agriculture.
http://soils.usda.gov

Open Directory Project: Earth Sciences
A comprehensive listing of internet resources grouped as Earth sciences, and categorized by subdisciplines.
http://dmoz.org/Science/Earth_Sciences/

Paleogeographic Atlas Project
Reconstructions of prehistoric maps of Earth's surface, from the Department of Paleogeography at the University of Chicago.
http://pgap.uchicago.edu

Physical Geography
Internet resources for the study of physical geography.
http://www.uwsp.edu/geo/internet/
    physical_geog_resources.html

The Geography Site: Physical Geography
An introduction for school students.
http://www.geography-site.co.uk/
    pages/physical.html

United States Geological Survey (USGS)
The federal portal for the science of Earth, its resources, and its hazards.
http://www.usgs.gov

University of California, Berkeley Library:
Earth Sciences Collection
Includes an extensive collection of maps, some available online.
http://www.lib.berkeley.edu/EART/

University of California, Berkeley: Museum of Paleontology: Plate tectonics
Provides animations that condense the geologic time of continental drift into a few seconds.
http://www.ucmp.berkeley.edu/geology/
    tectonics.html

USGS Geologic Time
An online publication explaining the science underlying concepts such as geologic time.
http://pubs.usgs.gov/gip/geotime/

Volcano World
A central reference point from the University of North Dakota for information on volcanoes worldwide.
http://volcano.und.nodak.edu

World Soil Information
A center of information about the world's soils and their importance for humanity.
http://www.isric.org

Worldwide Earthquake Locator
Daily updated reports and maps on earthquakes around the world from Edinburgh University Geography Department.
http://www.geo.ed.ac.uk/quakes/

INDEX